T0238971

IFIP Advances in Information and Communication Technology 435

IFIP – The International Federation for Information Processing

IFIP was founded in 1960 under the auspices of UNESCO, following the First World Computer Congress held in Paris the previous year. An umbrella organization for societies working in information processing, IFIP's aim is two-fold: to support information processing within its member countries and to encourage technology transfer to developing nations. As its mission statement clearly states,

> *IFIP's mission is to be the leading, truly international, apolitical organization which encourages and assists in the development, exploitation and application of information technology for the benefit of all people.*

IFIP is a non-profitmaking organization, run almost solely by 2500 volunteers. It operates through a number of technical committees, which organize events and publications. IFIP's events range from an international congress to local seminars, but the most important are:

- The IFIP World Computer Congress, held every second year;
- Open conferences;
- Working conferences.

The flagship event is the IFIP World Computer Congress, at which both invited and contributed papers are presented. Contributed papers are rigorously refereed and the rejection rate is high.

As with the Congress, participation in the open conferences is open to all and papers may be invited or submitted. Again, submitted papers are stringently refereed.

The working conferences are structured differently. They are usually run by a working group and attendance is small and by invitation only. Their purpose is to create an atmosphere conducive to innovation and development. Refereeing is also rigorous and papers are subjected to extensive group discussion.

Publications arising from IFIP events vary. The papers presented at the IFIP World Computer Congress and at open conferences are published as conference proceedings, while the results of the working conferences are often published as collections of selected and edited papers.

Any national society whose primary activity is about information processing may apply to become a full member of IFIP, although full membership is restricted to one society per country. Full members are entitled to vote at the annual General Assembly. National societies preferring a less committed involvement may apply for associate or corresponding membership. Associate members enjoy the same benefits as full members, but without voting rights. Corresponding members are not represented in IFIP bodies. Affiliated membership is open to non-national societies, and individual and honorary membership schemes are also offered.

Svetan Ratchev (Ed.)

Precision Assembly Technologies and Systems

7th IFIP WG 5.5
International Precision Assembly Seminar, IPAS 2014
Chamonix, France, February 16-18, 2014
Revised Selected Papers

 Springer

Volume Editor

Svetan Ratchev
The University of Nottingham
Institute for Advanced Manufacturing
University Park, Nottingham, NG7 2RD, UK
E-mail: svetan.ratchev@nottingham.ac.uk

ISSN 1868-4238   e-ISSN 1868-422X
ISBN 978-3-662-51626-3   e-ISBN 978-3-662-45586-9
DOI 10.1007/978-3-662-45586-9
Springer Heidelberg New York Dordrecht London

Typesetting: Camera-ready by author, data conversion by Scientific Publishing Services, Chennai, India

Printed on acid-free paper

Springer is part of Springer Science+Business Media (www.springer.com)

Foreword

The book includes a selected set of papers presented at the 7th International Precision Assembly Seminar (IPAS 2014) held in Chamonix, France, in February 2014. The International Precision Assembly Seminar, which was established in 2003 by the European Thematic Network Assembly-Net, has developed the premier international event for presenting and discussing the latest research, new innovative technologies, and industrial applications in the area of precision assembly.

We consider precision assembly as a process that covers a wide range of products where handling, positioning, manipulation, and joining technologies are constrained by specific quality, accuracy, and repeatability requirements that cannot be met by conventional assembly methods. At the micro-scale, a distinctive feature of precision assembly is that surface forces are often dominant over gravity forces, which determines a number of specific technical challenges including high-accuracy positioning and manipulation techniques, micro-gripping methods that take into account the surface forces, high-precision micro-feeding techniques, and micro-joining processes. At the other end of the scale, there are specific challenges in industries such as aerospace poised by the need to assemble large structures with extremely small tolerances that usually require additional metrology assistance and further equipment and process enhancement.

Precision assembly of complex high-value products is a key manufacturing process in sectors such as the automotive and aerospace sectors, and in defence, pharmaceutical, and medical industries. Some of the common trends underlining the development of precision assembly systems in these sectors include: increased demand for rapid ramp-up and downscale of production systems; increased demand for assembly systems that can react to disruptive events and fluctuations during the production process; and a drive toward after-sales service contracts for maintenance and equipment upgrade.

The book is structured into six chapters. Chapter 1 includes papers dedicated to micro-assembly processes and systems ranging from desktop factory automation and packaging of MEMS to self-assembly processes and platforms. Chapter 2 is focused on handling and manipulation and includes contributions on flexible gripper systems, fixturing, and high-precision actuators. Chapter 3 includes a range of contributions on tolerance management and error-compensation techniques applied at different scales of precision assembly. Chapter 4 describes some of the latest developments in metrology and quality control, while Chapter 5 introduces contributions on intelligent assembly control. Finally, Chapter 6 concludes with contributions on process selection, modelling, and planning.

The seminar is sponsored by the International Federation of Information Processing (IFIP) WG5.5, the International Academy of Production Research (CIRP), and the European Factory Automation Committee (EFAC). The seminar is supported by a number of ongoing research initiatives and projects including the European sub-technology platform in Micro and Nano Manufacturing MINAM 2.0, as well as the EU Framework 7-funded collaborative projects PRIME and COPERNICO.

The organizers should like to express their gratitude to the members of the international Advisory Committee for their support and guidance and to the authors of the papers for their original contributions. Special thanks go to Ruth Strickland and Rachel O'Shea from the Precision Manufacturing Centre at the University of Nottingham for handling the administrative aspects of the seminar, putting the proceedings together, and managing the detailed liaison with the authors and the publishers.

October 2014   Svetan Ratchev

Organization

Seminar Chair

Svetan Ratchev | Director of the Institute for Advanced Manufacturing, University of Nottingham, UK

Keynote Speakers

Danick Bionda | General Secretary of Micronarc, Switzerland

Irene Fassi | Head of Micro Enabled Devices and Systems Research Unit, ITIA-CNR, Institute of Industrial Technologies and Automation, National Research Council, Milan, Italy

James Kell | On-Platform Repair Technology Specialist, Rolls-Royce, UK

Benedetto Vigna | Executive Vice President, General Manager of the Analog, MEMS and Sensors Group of STMicroelectronics, Switzerland

Session Chairs

Markus Dickerhof | The Karlsruhe Institute of Technology, Germany

Jacques Jacot | EPFL, Switzerland

Pierre Lambert | EPFL, Switzerland

Richard Leach | National Physical Laboratory, Middlesex, UK

Timo Prusi | Tampere University of Technology, Finland

Erik Puik | Hogeschool, The Netherlands

Svetan Ratchev | University of Nottingham, UK

Alexander Steinecker | CSEM, Switzerland

Local Organizing Committee

Evelyne Roudier-Poirot | Manager of the Majestic Congress Centre and Convention Bureau of Chamonix, France

Conference Administration

Rachel O'Shea	Project Administrator, University of Nottingham, UK
Ruth Strickland	Project Administrator, University of Nottingham, UK

International Advisory Committee

T. Arai	University of Tokyo, Japan
H. Afsarmanesh	University of Amsterdam, The Netherlands
D Axinte	University of Nottingham, UK
M. Bjorkman	Linkoping Institute of Technology, Sweden
H. Bley	University of Saarland, Germany
D. Branson	University of Nottingham, UK
L.M. Camarinho-Matos	Universidade Nova, Portugal
J. Claverley	National Physical Laboratory, UK
A. Delchambre	ULB, Belgium
M. Desmulliez	Heriot-Watt University, UK
S. Dimov	University of Birmingham, UK
G. Dini	Univ di Pisa, Italy
S. Durante	DIAD, Italy
K. Ehmann	Northwestern University, USA
Irene Fassi	ITIA-CNR, Italy
R.W. Grubbstrom	Linkoping Institute of Technology, Sweden
T. Hasegawa	National College of Technology, Japan
H. Krieger	CSEM, Switzerland
P. Lambert	ULB, Belgium
R. Leach	National Physical Laboratory, UK
N. Lohse	University of Loughborough, UK
P. Lutz	LAB, France
H. Maekawa	NIAI Science and Technology, Japan
B. Nelson	ETH, Switzerland
J. Ni	University of Michigan, USA
D. Pham	University of Birmingham, UK
M. Pillet	Polytech Savoie, France
G. Putnik	University of Minho, Portugal
B. Raucent	UCL, Belgium
K. Ridgway	Sheffield University
K. Saitou	University of Michigan, USA
J. Segal	University of Nottingham, UK
W. Shen	National Research Council of Canada

M. Summers	Airbus, UK
J. Heilala	VTT, Finland
M. Tichem	TU Delft, The Netherlands
J. Jacot	EPFL, Switzerland
R. Tuokko	TUT, Finland
P. Kinnell	University of Nottingham, UK
E. Westkamer	Fraunhofer IPA, Germany
S. Koelemeijer	Jaeger-Lecoultre, Switzerland
D. Williams	Loughborough University, UK

Sponsoring Organizations

The International Academy for Production Engineering

International Federation for Information Processing   ifip

Table of Contents

Robust Adhesive Precision Bonding
in Automated Assembly Cells

Tobias Müller[1,*], Sebastian Haag[1], Thomas Bastuck[1], Thomas Gisler[2],
Hansruedi Moser[2], Petteri Uusimaa[3], Christoph Axt[4], and Christian Brecher[1]

[1] Fraunhofer Institute for Production Technology IPT, Steinbachstr. 17,
52074 Aachen, Germany
tobias.mueller@ipt.fraunhofer.de
[2] FISBA OPTIK AG, Rorschacher Str. 268, 9016 St. Gallen, Switzerland
[3] Modulight Inc., Hermaniaku 22, 33720 Tampere, Finland
[4] Rohwedder Micro Assembly GmbH, Opelstr. 1, 68789 St. Leon-Rot, Germany

Abstract. The assembly of optical components goes along with highest
requirements regarding assembly precision. Laser products have become an
integral part of many industrial, medical, and consumer applications and their
relevance will increase significantly in the years to come. Still economic challenges remain. Assembly costs are driven by the demanding requirements regarding alignment and adhesive bonding. Especially challenging in precision
bonding are the interdependencies between alignment and bonding. Multiple
components need to be aligned within smallest spatial and angular tolerances in
submicron order of magnitude. A major challenge in adhesive bonding is the
fact that the bonding process is irreversible. Accordingly, the first bonding attempt needs to be successful. Today's UV-curing adhesives inherit shrinkage
effects during curing which are crucial for the submicron tolerances of e.g.
FACs or beam combiners what makes the bonding of these components very
delicate assembly tasks. However, the shrinkage of UV-curing adhesives is not
only varying between different loads due to fluctuations in raw materials, it is
also changing along the storage period. An answer to this specific challenge can
be the characterization of the adhesive on a daily basis. The characterization before application of the adhesive is necessary for precision optics assembly in
order to reach highest output yields, minimal tolerances and ideal beam-shaping
results. The work presented in this paper aims for a significantly reduced impact
of shrinkage effects during curing of highly durable UV-curing epoxy adhesives
resulting in increased precision. Key approach is the highly precise volumetric
dispensing of the adhesive as well as the characterization of the shrinkage level.
These two key factors allow most reproducible adhesive bonding in automated
assembly cells. These proceedings are essential for standardized automated assembly solutions which will prospectively play a major role in laser technology.

1   Introduction

The assembly of optical components goes along with the highest requirements regarding assembly precision in many cases. For instance laser products have become an

---

* Corresponding author: Tel: +49-241-8904-493, Fax: +49-241-8904-6-493.

S. Ratchev (Ed.): IPAS 2014, IFIP AICT 435, pp. 1–7, 2014.
© IFIP International Federation for Information Processing 2014

integral part of many industrial, medical, and consumer applications and their relevance will increase significantly in the years to come. Still, economic challenges of laser related products prevent the breakthrough in many applications. Assembly costs are driven mainly by the demanding requirements regarding alignment and adhesive bonding in all spatial dimensions. Extensive research efforts are being made in matters of novel bonding methods, whilst the challenges of the state-of the art UV-curing adhesive bonding still provide opportunities for improvement as identified in several recent research projects and contributions ([1],[2],[3],[4],[5],[6]).

The major and most appreciated advantages of UV-curing adhesive bonding are its low temperature, the automation friendliness, and the reasonable costs for automation solutions. Yet, especially challenging in precision assembly are the interdependencies between alignment and adhesive bonding. Shrinkage effects during curing of the adhesives are crucial for the submicron tolerances of for instance fast-axis collimators (FACs). Also beam combiners are a very delicate assembly task. Multiple components sharing interdependencies need to be aligned within smallest tolerances in submicron order of magnitude. For such assembly tasks the major challenge in adhesive bonding at highest precision level is the fact, that the bonding process is irreversible. Accordingly, the first bonding attempt needs to be successful as especially for automated solutions dissolving the link between two components is not possible. The impact of the shrinkage effects can be tackled both by a suitable design of the bonding area and a positioning offset of the optic for compensation purposes. Yet, compensating shrinkage effects is difficult, as the shrinkage of UV-curing adhesives is not necessarily constant between two different loads due to fluctuations in raw materials and variations over the storage period even under ideal circumstances. An up-to-date characterization of the adhesive is necessary for automation in optics assembly to reach highest output yields, minimal tolerances and ideal beam-shaping results. Accordingly, today the operator needs to adjust the compensation offset data on a daily basis during the first assembly processes of the production series as practice shows. Chances of creating sub-standard goods are high for these first systems, which is problematic for high-value precision optics.

2   Geometric Model of the Bonding Area

The work presented in this paper aims for a significantly reduced impact of shrinkage effects of UV-curing adhesives and a resulting increase in precision in automated assembly cells. Key approach is the highly precise volumetric dispensing of the adhesive as well as an up-to-date characterization of the shrinkage level. These two key factors allow reproducible adhesive precision bonding in automated assembly cells.

First step in the approach for increased precision in adhesive bonding is to model the bonding system and identify and separate the origins for the misalignment effects in order to be able to compensate the shrinkage effects more efficiently. The model of the bonding area that was elaborated presumes that the misalignment is mainly caused by two effects: A spatial and an angular offset. Figure 1 shows the model of the bonding area.

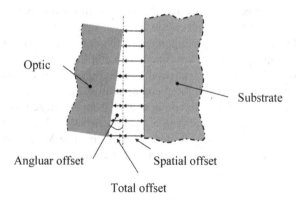

Optic

Substrate

Angluar offset   Spatial offset

Total offset

Fig. 1. Model for compensation of shrinkage effects

Spatially, the bonding area consists of an offset part which is causing a linear shrinkage in normal direction with reference to the substrate surface. This shrinkage effect is less crucial, as kinematically it leads to small movements which can be compensated linear. The bigger impact on the bonding area has the second wedge-shaped part as it leads to angular misalignments causing an angular (β) and lateral (Δx) displacement of the component due to the kinematic shown in Figure 2.

Fig. 2. Kinematics for misalignment caused by wedge-shaped bonding area

For optical components these angular (tip-tilt) misalignments are in many cases the most sensitive degrees of freedom.

3 Experimental Quantification of Adhesive Shrinkage

In order to determine the volumetric shrinkage during curing of the adhesive a measurement setup was built. The measurement principle is depicted in Figure 3. In order to quantify shrinkage effects a defined angle between the two bonding partners was set up. This predefined misalignment causes a wedge shaped gluing area. As the glue is cured, it shrinks and causes a misalignment of the optical component. The motion of the glass cube is monitored and measured by a PSD. Essential in order to create reliable and reproducible measurement results is the precise verification of the angle β. Therefore telecentric camera was setup and carefully aligned to the testing setup in order to measure the edges of the two components. Using an image processing routine the angle could be precisely set utilizing a micromanipulator holding the glasscube and manipulating it in the necessary degrees of freedom. The Micromanipulator used was the flexure based Commander 6 which has been developed at Fraunhofer IPT.

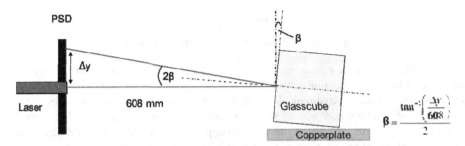

Fig. 3. Principle measurement setup for shrinkage quantification

With this measurement setup it was possible to determine angular movements with a precision of less than 10µrad covering a measurement range of up to 10°. Figure 4 shows an exemplary measurement of the shrinkage effects. The wedge angle in this measurement was 4.36° the spatial offset of 100µm.

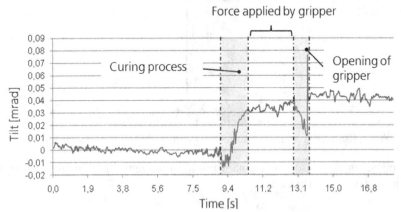

Fig. 4. Tilt measurement sample of curing process

By using a highspeed-PSD device it was possible to acquire the motion data in realtime which provides information about the temporal shrinkage behaviour during the curing process. Remarkable is a minor expansion at the beginning of the curing process followed by a fast expansion until the final curing result is reached. Opening the mechanical gripper leads to force amplitudes in both directions and a final relaxation leading to an increase of the tilt by about 10μrad. The quantification and variation of the gripper's closing force allowed an approximation of the forces induced by the adhesive during curing process. This knowledge is essential for bonding process development when the components shall be kept in place by the application of a retaining force. Therefore the gripper was actuated pneumatically allowing a precise an easy variation of the clamping force.

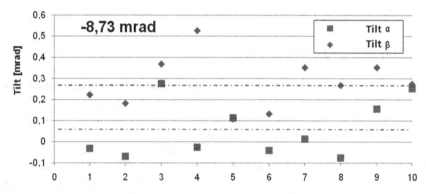

Fig. 5. Tilt measurement sample of curing process

Representative results for a measurement of one sample adhesive are shown in Figure 5. The initial wedge angle is this specific case was 8.73mrad (-0,5°). The mean tilt resulting from curing in the designated shrink direction (Tilt β) was at 0.287 mrad. The undesired parasitic tilt movement (Tilt α) was at 0.0596 mrad which represents the measuring uncertainty. Formula (1) is describing the linear shrinkage neglecting the linear shrinkage resulting from parallel movements of the components.

$$\Delta\alpha = -\alpha \cdot lS$$

$\Delta\alpha$ = Tilt induced by shrinkage

α = Adjusted wedge-angle

lS = linear shrinkage

The corresponding shrinkage in this experiment was 2.31% which is roughly double the value given in the technical data sheet. This deviation is caused by ageing effects of the adhesives during their shelf life. Over the time the UV-curable adhesives partially react even without any exposure to UV-light. Accordingly, the magnitude of shrinkage drops with the age of the adhesive.

However, the severity of shrinkage is not the only factor influencing the achievable bonding results. Also highly precise volumetric dosing and minimal adhesive volumes minimize the shrinkage effects. Therefore a qualification methodology for dosing systems has been elaborated allowing an efficient selection of the best suitable dosing system. Evaluation criteria are minimum dispensing volume, reproducibility, handling and costs. Jet dosing systems prove to be most reliable regarding dispensing volume reproducibility for the assembly of micro-optical components. A further increase in precision of volumetric dispensing could be achieved by creating minimal dropsizes and applying several drops to generate one drop of the desired volume. Once again, a telecentric camera and an image processing routine were used to characterize the reproducibility of the drop diameter. Based on these results a number of suitable adhesives for bonding optics has been validated and characterized regarding shrinkage developments over their lifetime.

4   Conclusion and Outlook

Result of the work presented in the paper is the precise predictive compensation of shrinkage effects of UV-curing adhesives during the curing process. Firstly, the importance of precise volumetric dosing was elaborated. Furthermore the impact of shrinkage effects has been modelled, described and quantified. The determination of the volumetric shrinkage magnitude is possible with the measurement setup described. The shrinkage could be determined as precisely a 0.1 %. During the experimental phase shifts in shrinkage of the same adhesive of more than 5% due to ageing effects were observed. For precise and efficient compensation of the shrinkage effects a measurement of the shrinkage magnitude instantly before the glue application appears to be unavoidable. The determination of the shrinkage value is even more important in automated assembly solutions, as the offset values can thus be calculated dynamically.

Looking ahead automation will play a growingly important role in laser optics assembly. For robust automation solutions in optics assembly producing best results the detailed knowledge of the adhesives behaviour as well as a sophisticated compensation of the shrinkage is obligatory. Accordingly, the proceedings presented in this paper are essential for standardized automated assembly solutions.

Acknowledgement. Fraunhofer IPT would like to thank the EC for supporting parts of this work within the APACOS project in the 7th Framework Program.

References

1. Beckert, E., Burkhardt, T., Eberhardt, R., Tünnermann, A.: Solder Bumping – a flexible Joining Approach for the Precision Assembly of optoelectronical Systems. In: Micro-Assembly Technologies and Applications. IFIP Tc5 Wg5.5 Fourth International Precision Assembly Seminar (IPAS 2008), Chamonix, France, February 10-13, pp. 139–148. Springer (2008)
2. Miesner, J., Timmermann, A., Meinschien, J., Neumann, B., Wright, S., Tekin, T., Schröder, H., Westphalen, T., Frischkorn, F.: Automated Assembly of Fast Axis Collimation (FAC) Lenses for Diode Laser Bar Modules Photonics West 2009, San Jose, January 26. PW09L-LA107-7198-15 Wash: Proc. SPIE 7198, 71980G (2009), doi:10.1117/12.809190

3. Brecher, C., Pyschny, N., Haag, S., Guerrero Lule, V.: Automated alignment of optical components for high-power diode lasers. In: Zediker, M.S. (ed.) High-power diode laser technology and applications X, San Francisco, California, United States, January 22-24. Proceedings of SPIE, vol. 8241, pp. 82410D–82410D-11. SPIE, Bellingham (2012)
4. Brecher, C., Pyschny, N., Haag, S., Mueller, T.: Automated assembly of VECSEL components. In: Hastie, J.E. (ed.) Vertical external cavity surface emitting lasers (VECSELs) III, San Francisco, California, United States, February 3-5. Proceedings of SPIE, vol. 8606, pp. 86060I–86060I-14 (2013)
5. Garlich, T., Guerrero, V., Hoppen, M., Müller, T., Pont, P., Pyschny, N., et al.: SCALAB. Scalable Automation for Emerging Lab Production. Final report of the MNT-ERA.net research project, 1. ed. Apprimus-Verl, Aachen (2012)
6. Pierer, J., Lützelschwab, M., Grossmann, S., Spinola Durante, G., Bosshard, C., Valk, B., et al.: In: Zediker, M.S. (ed.) High-power diode laser technology and applications IX, San Francisco, California, United States, January 23-25, vol. 7918, pp. 79180I–79180I-8. SPIE, Bellingham (2011)

Assembly of Silicon Micro-parts with Steel Spindles Using Low-Temperature Soldering

Laurenz Notter and Jacques Jacot

Laboratoire de Production Microtechnique (LPM)
École polytechnique fédérale de Lausanne (EPFL)
CH-1015 Lausanne, Suisse
laurenz.notter@a3.epfl.ch
http://lpm.epfl.ch

Abstract. A new assembly method for silicon micro-parts with non-planar steel parts is proposed: low-temperature soldering. Existing techniques for micro-mechanical assemblies are analyzed and compared to the proposed method, the method is explained and validated on an existing product using functional tests. Performances of the reference method (adhesive bonding) are not yet fully reached, but the results are close and partially fulfil the specifications of the current production.

Keywords: Assembly, micro-part, silicon, steel, low-temperature soldering, surface tension, micro-mechanics.

1   Introduction

The potential for producing complex micro-mechanical planar pieces from silicon using clean-room fabrication processes has been known for some time. In the field of micro-mechanics – we consider characteristic dimensions between 100 µm and 1 mm and dimensional precision in the micrometer-range – such silicon pieces can be very useful because of their high mechanical strength, their low geometric tolerances, low weight and the parallel fabrication method. Even though the production of silicon micro-pieces is possible using currently available technologies, their application is limited due to three main problems: the maximum achievable thickness, the limited feasibility of multi-layer designs and their integration into out-of-plane assemblies.

In this paper we describe an assembly method that permits to circumvent some of these limitations and compare it to existing techniques using functional tests.

Context and Motivation. Our analysis is part of a research project with industrial partners active in the watch industry and is funded by the Swiss government[1]. The research is guided by the need for a combination of silicon micro-parts with steel pieces so as to permit an integration into existing assemblies while keeping the costs at a viable level for an industrial production. For this reason, we focus on a typical

---

[1]   Commission for Technology and Innovation CTI.

S. Ratchev (Ed.): IPAS 2014, IFIP AICT 435, pp. 8–14, 2014.

assembly problem in this domain: the escape wheel and its spindle. In **Fig. 1** one of the realized assemblies is pictured.

Differentiation from MEMS. Unlike the widely researched field of MEMS, which operate on smaller scales, this paper focuses on the integration of silicon micro-parts into existing assemblies such as mechanical movements that are on the scale of some centimeters.

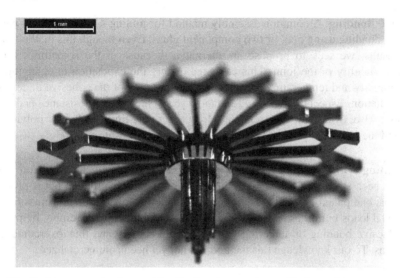

Fig. 1. An escape wheel assembly after low-temperature soldering and torque test

Fabrication of High-precision Mechanical Parts Made of Silicon. Clean-room fabrication processes such as (deep) reactive ion etching (RIE or DRIE) have been used to produce high-precision micro-mechanical parts from silicon wafers such as gears, springs [2,5] and more complex pieces such as multi-level components [3] or rotors for micromotors [4].

Compared to metal parts fabricated using techniques such as fine-blanking, mechanical silicon parts offer greater liberty for the design of their shape while guaranteeing sub-micrometer dimensional precision. As silicon has a lower density and a higher tensile strength than stainless steel (see [6]), mechanical applications benefit from the possibility of lightweight yet resistant designs.

2   Established Assembly Methods for Silicon Micro-parts with other Materials

The integration of mechanical silicon parts into assemblies using other materials such as steel or nickel is still a challenge. Conventional (macroscopic) mechanical assembly methods rely heavily on properties common for most metals such as plastic deformation (screwing, press-fitting) thermal diffusion and dilatation (soldering, welding, riveting). As these properties are weak or non-existent for silicon and for its native oxide, most conventional techniques are not easily applicable.

In this study we focus on the combination of planar silicon micro-parts with metallic out-of-plane parts such as spindles and compare the established assembly technique to our proposed method of low-temperature soldering.

Adhesive Bonding. A common assembly method for joining silicon and steel parts is adhesive bonding using one- or two component glues. Even though this method yields good results, we seek to propose an alternative because of the sometimes limited chemical stability of the joint. This concerns mainly the exposition to changing ambient humidity and temperature as well as the degradation of the adhesive under ultraviolet radiation. Especially the long term (>10 years) mechanical resistance is difficult to assess. Also, cycle times are relatively long due to the time needed for polymerization[2] and the assembly process is not easy to get under control.

Press-fitting with Compliant Structures. Press-fitting of structured silicon microparts has been shown to be a viable solution to our assembly problem but is protected by at least one patent [1]. This solution is applicable for low torque transmissions and small axial loads but it has also been proposed to combine a pre-assembly by press-fit with adhesive bonding or soldering to "lock" the assembly and thus overcome these limitations. To our knowledge this method has not yet been commercialized.

Low-temperature Soldering. Low-temperature soldering of silicon to steel is a technique that is very similar in application to adhesive bonding (when a convection oven and solder paste is used). Some of the drawbacks of adhesive bonding can be avoided with low-temperature soldering as soldered joints exhibit generally higher chemical stability as well as a resistance to humidity and ultraviolet radiation.

2.1   Comparison of Methods

We propose a qualitative overview of the above mentioned assembly methods that is based on criteria established by a functional analysis of the escapements of mechanical movements, as they are widely used in the Swiss watch industry. In addition to the precise geometric tolerances and mechanical resistance, a long lifespan is required. As the materials of the pieces to assemble are inert, the lifespan is limited by the stability of the joint. Wear may be caused by frequent exposure of the product to rapidly changing ambient atmospheric conditions such as temperature, UV radiation and humidity.

---

[2]   Usually in the range of some minutes to several hours, depending on the type of adhesive.

Table 1. Overview of assembly methods for silicon wheels with steel spindles

Criteria	Adhesive bonding	Press-fitting with compliant structures	Low-temperature soldering
Prototyping possible?	yes	no	(yes)[3]
Mechanical resistance	medium	low	medium
Chemical stability	medium	high	high
Disassembly possible?	no	yes	no
Cycle time	medium	short	short

Given the qualitative overview of **Table 1** it is reasonable to explore the potential of low-temperature soldering. We expect similar performances to adhesive bonding while potentially improving the long-term reliability of the assembly and reducing costs due to shorter cycle times.

3   Assembly of Silicon Escape Wheels with Their Spindles

One of the assemblies realized using low-temperature soldering is depicted in **Fig. 1** and its schematic representation can be found in **Fig. 2**. To qualify the result of our proposed method we compare our samples to an existing product that is assembled by adhesive bonding: the escape wheel with its steel spindle. We also use the same tests and criteria, which are used to qualify the ongoing production, for evaluating our assembly method.

The angular deviation from perpendicularity β and the concentricity error d are used to qualify the geometry, whereas the mechanical resistance is characterized by the maximum transmissible torque T_{max}. The visual aspect is very important for this type of assembly, but its objective measurement goes beyond the scope of this paper.

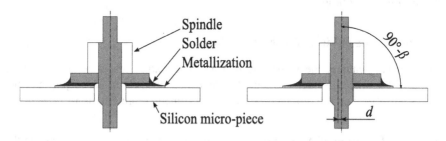

Fig. 2. Schematic representation of the assembled escape wheel with an indication of the concentricity error (distance d) and the deviation from perpendicularity (angle β)

---

[3] A metallization layer can easily be applied to the whole surface of a piece, but as soon as localized metallization is needed, prototyping is not possible anymore.

Preliminary Condition for Soldering of Silicon Micro-parts. Silicon forms very stable oxides in ambient atmosphere that cannot be removed using simple cleaning processes or fluxes. A convenient solution is to deposit a metallization Ti-Ni-Au[4] on the silicon substrate, so that we can rely on decades of experience in assembly of metallic surfaces by soldering.

Silicon oxides are not wetted by molten metal, this implies that a localized metallization can be used to control the precise location of the solder joint. Depending on the geometry of the pieces to assemble, this property can be used to hide the solder joint.

4   Soldering Process

As is succinctly described by Humpston and Jacobson: *"Soldering and brazing involve using a molten filler metal to wet the mating surfaces of a joint, with or without the aid of a fluxing agent, leading to the formation of metallurgical bonds between the filler and the respective components. In these processes, the original surfaces of the components are 'eroded' by virtue of the reaction occurring between the molten filler metal and the solid components, but the extent of this 'erosion' is usually at the microscopic level (<100 µm, or 4000 µin.)."* [7]

In this section we retrace the different steps of the process and discuss particularities of our application. The usual preconditions for solder joints apply: the surfaces to join should permit a wetting of the filler metal. During the soldering the temperatures of the pieces and their mating surfaces must stay above the liquidus of the filler metal.

4.1   Degreasing before Soldering

A cleaning (degreasing) of the pieces is often necessary before the soldering takes place. The escape wheels come from the cleanroom and therefore they are already as clean as they can reasonably get. The spindles have been cleaned using an ultrasonic bath with filtered isopropanol.

4.2   Positioning of Pieces for the Assembly

Following the cleaning, the spindle is inserted in a fixture that assures its relative position to the wheel. Solder and flux are deposited on the surface to join and the wheel is positioned on top of it (see **Fig. 2** for a schematic representation).

Mechanical Precision of Assembly. We use two geometrical criteria to evaluate the functionality of the assembly: the perpendicularity and the concentricity between spindle and wheel. Depending on the design of the spindle, perpendicularity can be achieved "by default" as the surface tensions of the liquid solder will equilibrate the

---

[4]   Titanium is used for adhesion on the silicon substrate, the nickel-layer serves as interface for the solder joint, whereas the gold is applied on top to prevent premature oxidation of the underlying nickel.

distance between the metallized silicon and the mating surface of the steel spindle. The concentricity error is limited by the clearance between the diameter of the spindle and the inner diameter of the wheel.

Quantity of Metal for Solder Joint. In assemblies of this scale it is not trivial to deposit the optimum quantity of filler metal at the joint. During our research we have qualified several deposition methods but finally the most basic ones – solder paste dispensing and solder preforms – have proven to be sufficient for this application.

4.3   Soldering Operation

The fixture with the pre-assembled pieces is heated using a convection oven with control of the temperature profile. We focused on low-temperature solder alloys so that temperatures stay below 200°C, thereby avoiding any coloration of steel that may happen at or above that point due to additional surface oxidation.

Removal of Surface Oxides. Chemical agents that remove surface oxides (fluxes) are commonly used. Sometimes they are complemented by an inert atmosphere for reducing supplementary surface oxidation during the heating, for example using nitrogen. In our case, the removal of surface oxides concerns mainly the steel piece and the filler metal itself.

4.4   Cleaning after Soldering

Depending on the type of flux used, the subsequent cleaning-process is more or less complex. The use of water-soluble fluxes has been found to be sufficient and therefore cleaning is done in aqueous ultrasonic baths.

5   Results

In section 3 we explained the design of the escape wheel assembly that has been used for comparative tests. In **Table 2** the geometrical errors β (deviation from perpendicularity) and d (concentricity) as well as the maximum transmissible torque T_{max} are reported for two assembly methods. Target values are also reproduced to allow a more comprehensive comparison.

Table 2. Comparison of two methods by measuring the performances of the assemblies. For every variable the mean value and the estimator of the standard deviation are indicated.

Method	$\bar{\beta}$	$s(\bar{\beta})$	$\bar{d}$	$s(\bar{d})$	$\bar{T}_{max}$	$s(\bar{T}_{max})$
Adhesive bonding	0.3°	0.18°	6 μm	1.8 μm	4.5 mNm	0.6 mNm
Low-temperature soldering	0.7°	0.45°	6 μm	5.1 μm	3.4 mNm	1.2 mNm
Target value	<0.5°	-	<12 μm	-	>2 mNm	-

Discussion of Results. Our proposed alternative approach does not yet fully reach the performances of the current assembly method, however the results are close enough to be very promising. Two out of three criteria are already above the acceptance limits of the ongoing production, and we have evidence that a significant improvement for the third one is possible.

6   Conclusion

The presented approach of low-temperature soldering for the assembly of silicon micro-parts with steel spindles is a valid alternative to conventional adhesive bonding. The comparison with the current assembly method shows that the performances are close enough to lead to an equivalence of the methods in the near future and that an industrialization is of interest.

Identified Subjects for Future Research. Long term comparative tests between adhesive bonding and low-temperature soldering should be carried out, notably concerning the mechanical resistance. Also, it seems that the formation of intermetallic compounds in the boundary layers has a major influence on the transmissible efforts and the resilience to mechanical shocks. Therefore it would be interesting to characterize mechanical performances in function of used alloys, fluxes and microstructure of the interfaces.

Acknowledgements. We thank our partners Mimotec SA and Sigatec SA as well as the members of our project team for their efforts and the pleasant cooperation.

References

1. Conus, T., Verardo, M., Kohler, F., Saglini, I., Jurin, A.C., Jacot, J., Perret-Gentil, M.: Assembly of a part not comprising a plastic range (2012)
2. Guckel, H., Skrobis, K.J., Christenson, T.R., Klein, J., Han, S., Choi, B., Lovell, E.G.: Fabrication of assembled micromechanical components via deep x-ray lithography, pp. 74–79 (1991)
3. Huber, R., Conrad, J., Schmitt, L., Hecker, K., Scheurer, J., Weber, M.: Fabrication of multilevel silicon structures by anisotropic deep silicon etching. Microelectronic Engineering 67–68, 410–416 (2003); Proceedings of the 28th International Conference on Micro- and Nano-Engineering
4. Mehregany, Gabriel, K., Trimmer, W.: Micro gears and turbines etched from silicon. Sensors and Actuators 12(4), 341–348 (1987)
5. Um, D.: Micro scale silicon dioxide gear fabrication by bulk micromachining process (2010)
6. Petersen, K.E.: Silicon as a mechanical material. Proceedings of the IEEE 70(5), 420–457 (1982)
7. Humpston, G., Jacobson, D.M.: Principles of Soldering. ASM International (2004)

Testing the Mechanical Characteristics and Contacting Behaviour of Novel Manufactured and Assembled Sphere-Tipped Styli for Micro-CMM Probes[*]

Dong-Yea Sheu[1], James D. Claverley[2], and Richard K. Leach[2]

[1] Dept. of Mechanical Engineering, National Taipei University of Technology
1, Secc.3, Chung-Hsiao E. Rd. Taipei 106, Taiwan
dongyea@ntut.edu.tw
[2] Engineering Measurement Division, National Physical Laboratory
Hampton Road, Teddington, Middlesex, TW11 0LW, UK
James.claverley@npl.co.uk

Abstract. The effectiveness and versatility of the probes used on micro-co-ordinate measuring machines is currently limited by the dimensions, geometry and quality of the spherical stylus tips available. A shaft fabrication and gluing process is demonstrated to assemble a sphere-tipped stylus for tactile micro-co-ordinate measuring machine probes. The contact behaviour of the manufactured micro-styli is investigated both in a static and a vibrating mode. The ultimate strength of the glue joint is also investigated. It is concluded that the assembly of styli using the presented method is viable, however, that stylus shafts below 40 µm diameter may require new manufacturing methods.

Keywords: micro-CMM, sphere tipped stylus, assembly, gluing process, WEDG, micro-EDM.

1   Introduction

Geometrical measurement of micro-components is a constant issue for micro-manufacturing engineers. Although several optical techniques are available, some geometries, such as holes, channels and surfaces with high curvature, still need to be characterised using a tactile micro-co-ordinate measuring machine (micro-CMM). To continue to achieve accurate results while using tactile micro-CMMs, it is essential to fabricate high quality sphere-tipped micro-styli.

Using a combination of wire electro discharge grinding (WEDG) [1] and one-pulse electro discharge (OPED) technology [2], sphere-tipped micro-styli can be fabricated with spherical tip diameter of approximately 0.07 mm [3]. During the OPED process, an instant electro-discharge is focused onto the end of a stylus shaft. This process

---

[*] © Crown copyright 2014. Reproduced by permission of the Controller of HMSO and the Queen's printer for Scotland.

S. Ratchev (Ed.): IPAS 2014, IFIP AICT 435, pp. 15–21, 2014.
© IFIP International Federation for Information Processing 2014

results in the formation of a spherical tip on the stylus shaft, creating a monolithic sphere-tipped micro-stylus. However, the instant electro-discharge energy shock can result in a spherical tip with a high sphericity error, in the order of 500 nm to 800 nm [2]. To address this, a hybrid WEDG and gluing process was developed to assemble the stylus tips. The manufacturing and positioning capability of the WEDG machine are combined to achieve high accuracy assembly. Following assembly, it is essential that the glue strength is determined to ensure the manufactured styli are suitable for use with existing micro-CMM probes. This paper describes the results of an investigation into the glue strength of the manufactured sphere-tipped styli. The experimental process used to investigate the triggering behaviour of probes using these assembled styli in static and dynamic (vibrating) modes is also described.

2   Sphere-Tipped Micro-stylus Assembly

Using a hybrid WEDG and gluing process, micro-styli with sphere tips of diameter approximately 0.07 mm were assembled. This process is fully described in a previous study [4]. The shaft manufacture and assembly process (glue deposition, manipulation and assembly) was carried out on a single machine. Using a CCD camera for position assistance, it was possible to assemble the sphere-tipped stylus with an accuracy of approximately 1 μm. Once manufactured, the micro-shaft is dipped into a glue reservoir and then assembled with a sphere secured in a small vacuum holder. A schematic of the manufacturing and assembly process is shown in figure 1.

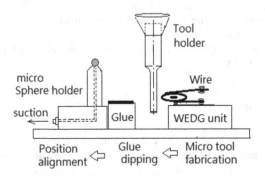

Fig. 1. Sphere-tipped micro-stylus manufacturing and assembly setup [4]

High aspect ratio sphere-tipped micro-styli can be assembled using this hybrid gluing and WEDG process. This is due to the high quality of straight and long shafts that are easily fabricated using the WEDG process. Two suitable designs for styli are shown in figure 2, along with the manufactured styli.

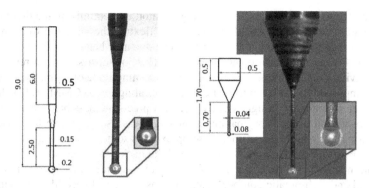

Fig. 2. Assembled micro-styli. Working aspect ratios are 12.5 (left) and 8.75 (right). All dimensions are in millimetres.

3   Mechanical Properties of the Manufactured Micro-styli

3.1   Contact Behaviour

An important parameter of micro-CMM probes is the contact force they impart on the measurement surface during probing. Although the stylus itself has some effect on this parameter, the probing system, especially its stiffness and mode of operation, is the major contributor. Therefore, only an indication of the effects apparent when probing at the micro-scale can be investigated using just the stylus, such as the effect of the surface interaction forces. The true probing force cannot be determined without installing the stylus onto a micro-CMM probe for testing.

When probing at the micrometre scale with a micro-CMM probe, the surface interaction forces have the effect of causing the probe to snap-in to the measurement surface on approach and/or to stick to the measurement surface after probing, causing snap-back [5]. To address this effect, the stylus tip should be vibrated. This technique is used in several newly developed micro-CMM probes [6] [7]. An experiment was designed to investigate the different contacting behaviours of the stylus tips during static and dynamic operation. A schematic of the experimental setup is shown in figure 3.

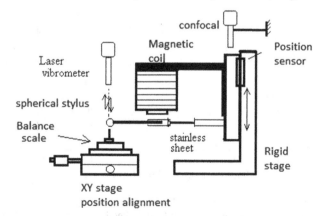

Fig. 3. Schematic of the dynamic contact force testing for the stylus

By using the magnetic coil as a vibration generator, and optimizing the drive current and the length and thickness of the stainless steel flexure, the vibration frequency and amplitude of the test stylus could be adjusted. The dynamic behaviour of the test stylus was recorded using a laser Doppler vibrometer. This low cost setup can result in the tests stylus vibrating at a frequency of 170 Hz and an amplitude of 1 µm. This resulting vibration amplitude is similar to that of existing vibrating micro-CMM probes; howeverer the vibration frequency is significantly lower. A precision mass balance was used as a force sensor, and a precision manipulation stage was used to move the stylus into contact during testing. The experimental setup exhibits some hysteresis due to the slow reaction time of the precision mass balance and the backlash on the screw-drive of the manipulation stage. To correct for this, the manipulation stage was operated at a slow travel speed and a chromatic confocal distance sensor, the CHRocodile E system from Precitec Optronik GmbH, was used to determine the true motion.

The contact behaviour of the stylus and the test measurement surface in static and dynamic mode was tested. The same setup was used in each case, as the stiffness of the steel flexure was determined to be very high in comparison with the stylus shaft, by at least three orders of magnitude. A set of representative results are shown in figure 4. It can be seen that, during static testing, the test stylus remained stuck to the measurement surface for approximately 1.5 µm post contact. This resulted in up to 1 µN of sticking force, which is usually attributed to the surface interaction forces [8]. During dynamic testing, with vibration characteristics as previously described, the same stylus exhibited no appreciable sticking to the surface while retracting.

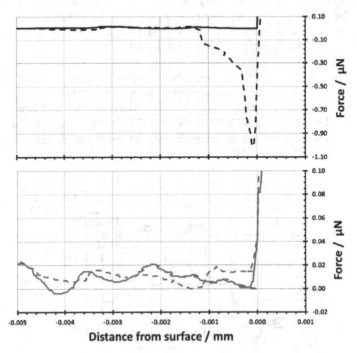

Fig. 4. The change in retract behaviour between static probing (top) and dynamic probing (bottom). The solid lines indicate approach, and the dashed lines indicate retract.

3.2   Glue Strength

The main use for these sphere-tipped micro-styli is as the contacting part of tactile micro-CMM probes [9]. For this application, knowledge of the stiffness of the stylus is required so that accurate determination of the transfer of mechanical contact to the sensing part of the CMM probe can be known. Also, the ultimate gluing strength should be determined to ensure there is no risk of stylus breaking during use. The gluing force of a glass spherical stylus tip has been previously measured to be in the order of 12 mN [4]. When testing a stylus of a similar design to that shown in figure 2 (left) but with a shaft diameter of 65 μm, a glue strength of 22 mN ± 1 mN ($k = 1$) was measured. A second stylus with a shaft diameter of 28 μm exhibited a glue strength of 5 mN ± 1 mN ($k = 1$), with an ultimate yield strength of the stylus shaft of 17 mN ± 1 mN ($k = 1$). Combined with the previously reported results for a 40 μm diameter of 12 mN, it is suggested that there is a first order linear relationship between the diameter of the stylus shaft and the glue strength of the assembled spherical tip. Due to the bespoke nature of these styli, the repeatability of the assembly process could not be determined. The experimental setup to complete these glue strength experiments was similar to that shown in figure 3, but with the test stylus held rigidly, rather than with a steel flexure.

3.3   Glue Strength – Dynamic Mode and Comparative Measurements

It was also possible to test the glue strength of the assembled stylus during dynamic operation. When operating in a dynamic mode, with a vibration amplitude of 1.2 μm and a vibration frequency of 161 Hz, a stylus with a shaft diameter of 40 μm showed a glue strength of 10 mN. This is in line with the previous results from the static glue tests.

A set of tests were also completed on styli manufactured using the previously described hybrid WEDG/OPED process. These monolithic styli exhibit very high strengths; a stylus with a shaft diameter of 40 μm showed an ultimate tensile strength of over 50 mN. The difference between the breaking modes of the assembled styli and the monolithic styli is shown in figure 5.

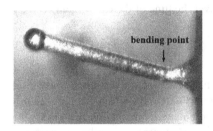 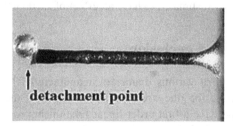

Fig. 5. Breaking of a WEDG/OPED monolithic (left) and a micro-assembled stylus (right)

However, with smaller shaft diameters becoming more prevalent, to accommodate smaller diameter sphere tips, WEDG technology becomes inappropriate for stylus shaft manufacture. This is due to the variability of the manufactured surface, which is not smooth, but instead is badly damaged and pitted. These rough features can act as initiation points for cracking, resulting in significantly weaker styli. For example, one tested stylus, manufactured using the hybrid WEDG/OPED process, with a shaft diameter of 40 µm, showed an ultimate tensile strength of only 6 mN, with the elastic limit being reached around 5 mN.

This issue of WEDG manufacture for small diameter shafts (below 50 µm) could be addressed through the use of electro-chemical machining (ECM). The surface roughness of a probe shaft manufactured using the ECM process is much lower than that formed by WEDG, which could result in stronger styli with diameters smaller than 40 µm. A trial micro-stylus with a tip diameter of 8 µm has been manufactured to demonstrate the capability of a hybrid ECM and OPED process. This stylus is shown in figure 6.

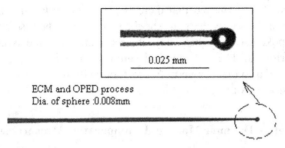

Fig. 6. A micro-stylus manufactured using a hybrid ECM and OPED process. The stylus tip has a diameter of 8 µm

3.4   Lifetime Investigation

To investigate the suitability of the assembled styli to perform as styli for micro-CMM probes, a short lifetime investigation was completed. An assembled probe, shaft diameter 40 µm, was contacted, in dynamic mode, to a force of 2 mN. This was repeated ninety times over 4.5 hours. During this time, no appreciable change in contact response was detected beyond that expected for the temperature fluctuations in the laboratory (estimated to have an effect of ± 200 nN over the time of the experiment).

4   Conclusion

Styli of various diameters, manufactured by this new assembly method, have been tested for glue strength. The strengths of the glue joints range from 5 mN to 22 mN, and have a first order linear relationship to the diameter of the stylus shaft. The glue strength results of the assembled styli have been compared to similar strength (the elastic limit) for monolithic styli and found to be significantly lower. This disparity is due to the inclusion of a glue joint in the assembled styli.

The strengths of the monolithic styli can be around 50 mN for styli diameters above 40 µm. The strength of the styli shafts manufactured using WEDG once the shaft diameter is below 40 µm is significantly lower. This is due to the high surface roughness produced during the WEDG process, whose features act as initiation points for cracking. It is suggested that the ECM process may be useful for the production of stylus shafts below 30 µm diameter. Styli shafts produced using ECM have lower surface roughness than those produced by WEDG, which reduces the chance of surface features acting as initiation points for cracking.

A low cost setup was designed and built to operate the test styli in a dynamic (or vibrating) mode. Through the use of a precision mass balance, the contact behaviour of the assembled styli was investigated. During static testing, the effect of the surface interaction forces resulted in the stylus sticking to the measurement surface with a force of approximately 1 µN. During dynamic testing, the same stylus exhibited no appreciable sticking to the surface while retracting. It is therefore concluded that dynamic operation is essential for micro-CMM probe operation at the micro-scale.

Acknowledgements. This work is funded by the UK National Measurement Office Engineering and Flow Metrology Programme 2011 to 2014 and also through EMRP Project IND59 – Microparts. The EMRP is jointly funded by the EMRP participating countries within EURAMET and the European Union.

References

[1] Masuzawa, T., Fujino, M., Kobayashi, K.: Wire electro-discharge grinding for micro-machining. CIRP Annals 34(1), 431–434 (1985)

[2] Sheu, D.-Y.: Study on an evaluation method of micro CMM spherical stylus tips by micro-EDM on-machine measurement. Journal of Micromechanics and Microengineering 20(7), 075003 (2010)

[3] Claverley, J.D., Sheu, D.-Y., Burisch, A.: Assembly of a novel MEMS-based 3D vibrating micro-scale co-ordinate measuring machine probe using desktop factory automation. In: Proc. IEEE ISAM (2011) ISBN 9781612843414

[4] Sheu, D.-Y., Cheng, C.: A Hybrid Microspherical Styli Gluing and Assembling Process on Micro-EDM. Materials and Manufacturing Processes (2012)

[5] Bos, E.J.C.: Aspects of tactile probing on the micro scale. Precision Engineering 35(2), 228–240 (2011)

[6] Claverley, J.D., Leach, R.K.: Development of a three-dimensional vibrating tactile probe for miniature CMMs. Precision Engineering 37(2), 491–499 (2013)

[7] Bauza, M.B., Hocken, R., Smith, S.T., Woody, S.C.: Development of a virtual probe tip with an application to high aspect ratio microscale features. Rev. Sci. Instrum. 76(9), 95–112 (2005)

[8] Claverley, J.D., Georgi, A., Leach, R.K.: Modelling the Interaction Forces between an Ideal Measurement Surface and the Stylus Tip of a Novel Vibrating Micro-scale CMM Probe. Precision Assembly Technologies and Systems, 131–138 (2010)

[9] Bosch, J. (ed.): Coordinate measuring machines and systems, First edit. Dekker, M. (1995)

Ultrasonic Press–Fitting:
A New Assembly Technique

Csaba Laurenczy, Damien Berlie, and Jacques Jacot

Laboratoire de Production Microtechnique (LPM)
École Polytechnique Fédérale de Lausanne (EPFL)
CH-1015 Lausanne, Suisse
Csaba.Laurenczy@alumni.epfl.ch

Abstract. The superposition of ultrasonic frequency vibrations to conventional machining techniques is known and practiced since the 1950s under the name of ultrasonic machining. Using ultrasound, many good properties appear including reduced thrust force, improved surface finish, reduced residual stress in machined material, etc… In this paper we present a new assembly technique based on the same principles: ultrasonic press–fitting. Feasibility and energy reduction are demonstrated through experiments under industrial production conditions.

Keywords: ultrasonic assembly, ultrasonic press–fitting, interference–fitting, thrust force reduction, hole–pin insertion energy reduction.

1   Purpose of this Paper

The vibrations assisted machining is an extension of conventional machining wherein a mechanical vibration is superposed on the tool movement. If the vibrations frequency exceeds 20 kHz, the expression mostly used in the literature is *ultrasound assisted machining*. This idea of assistance by ultrasound to the machining energy was also the starting point in the research conducted at EPFL – LPM. However, after several years of research, the authors came to the conclusion that the energy intake of ultrasound is higher by at least an order of magnitude than the one of the conventional machining or assembly techniques. Because the role of ultrasound surpasses the meaning of the word *assistance* usually used in the literature, the authors prefer the term of *ultrasonic machining* as well as *ultrasonic press–fitting* and will use them in this article instead of the usual expressions. Moreover, the mechanical behavior of the press–fitting as well as the hole–pin interactions are seen to be transformed in presence of ultrasound. Such a change in friction conditions and elastic–plastic characteristics have also been reported for other ultrasonic manufacturing processes [1].

The purpose of this paper is to present a new assembly technique: ultrasonic press–fitting. To achieve this goal, conventional press–fitting is introduced in section **2** and informations about ultrasonic equipment in section **3**. The description of the experimental setup of section **4** is followed by the measurement results in section **5** and their discussion in section **6**.

S. Ratchev (Ed.): IPAS 2014, IFIP AICT 435, pp. 22–29, 2014.

2   Conventional Press–Fitting

Process Steps. Press–fitting is a common assembly technique which consists in a fastening achieved by introducing a pin into a hole with an interference. Even though this process appears as simple, it is not straightforward at (sub)millimetric scale.

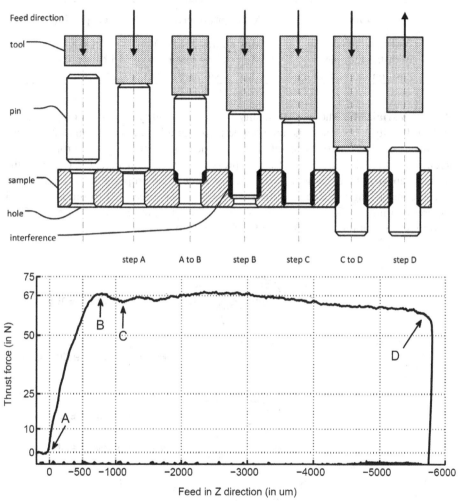

Fig. 1. Conventional press–fitting steps: **A** Beginning of the press–fitting with the contact between the tool and the pin resulting in an important thrust force increase **A to B** Feeding with growing pin–hole contact length resulting in a thrust force increase **B** Lower pin extremity reaches the minimal hole diameter **C** Lower pin extremity reaches the lower hole exit **C to D** Feeding with constant pin–hole contact length **D** Pin reaches its final position and tool moves back to its initial position **Process parameters.** CuZn$_{39}$Pb$_2$ brass sample, sample thickness 1000 µm, hole diameter 991 µm, 100Cr6 steel pin, pin diameter 998 µm, interference 7 µm, feed rate 10 mm/s.

Functional Analysis. Functions of the press–fitting are often observed to fail over the time. These cases have been studied especially through examples from the watchmaking industry by Bourgeois [2]. Some of the main functions expected from press–fitting are listed in **Table 1**.

Table 1. Result of a functional analysis modified from [3]

Nr.	Function	Acceptation criteria	Acceptation level
1	Withstand axial load	Min. sliding load (F)	$10\,N < F$
2	Withstand torque	Min. sliding torque (T)	$5\,N{\cdot}mm < T$
3	Position in axial direction	Max. position error (e)	$e < 2\,\mu m$

Process Variability. The lack of quality, i.e. the variability in process output, is due to the difficulty to maintain the hole diameter within tolerances of 1 μm to 2 μm during manufacturing. These extremely thin tolerances are necessary to control the interference and thus the thrust force during press–fitting. This statement is based on the Lamé–Clapeyron model which predicts thrust force to be proportional to interference and to contact length between parts. Hence, the thrust force used to run and control the press–fitting varies of 10 N to 50 N for a change in interference of 1 μm [2]. This represents a variability of 15 % to 75 % of the maximal thrust force.

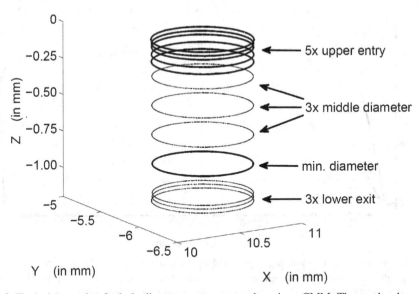

Fig. 2. Typical example of a hole diameter measurement by micro–CMM. The results showed that drilled–bored holes are not cylindrical as usually modeled and that a tightening of 1–4 μm exists at 50–100 μm above the lower hole exit. **Measure parameters.** 52 runs, probe diameter 302 μm, nominal hole diameter 990 μm.

Model of a Hole. To get a better understanding of the interference variability, the most accurate possible profile of 52 typical drilled–bored holes were measured at the Swiss Federal Institute of Metrology (*METAS*) on an enhanced micro–CMM on which a spherical probe with a diameter of 302 µm was mounted [4]. A representative result is plotted in **Fig. 2**. Each hole was measured at 12 different depth with more than 400 points per depth. The uncertainty of measurement in the three dimensions lies under 50 nm. The analysis of these measurements showed that drilled–bored holes are not cylindrical as usually modeled and that there is a tightening of 1 µm to 4 µm at 50 µm to 100 µm above the lower hole exit.

Diameter Measurement. Since all the samples cannot be sent for a micro–CMM measurement, the authors investigated the most common method used in watchmaking industry: gauges. An R&R test with 3 experienced operators, 2 measurements per hole and 13 holes was performed [5]. This test gave a repeatability σ_1 of 0.29 µm while the reproducibility σ_2 was 0.81 µm. The authors were not surprised to find out that as usually at (sub)millimetric scale the dispersion of the measuring instruments $\sigma_m = 0.85$ µm, was of the same order of magnitude that the dispersion of the hole manufacturing $\sigma_p = 0.89$ µm. Hence, the interference values should be taken with precautions.

3   Ultrasonic Equipment

Ultrasound Generation. Since the first ultrasonic experiments in the 1950s, the principle of operation of ultrasonic machining or ultrasonic welding systems remained identical [6, 7]. The ultrasonic press–fitting system is no different. A high frequency voltage generator converts a network supply voltage of 220 V at 50 Hz into a 1 kV voltage adjusted in frequency to the resonance frequency of the system. A second stage converts the electrical energy into longitudinal compression–tension mechanical vibrations by means of a piezoelectric transducer.

Ultrasound Amplification. The sonotrode, sometimes also designated by *horn*, *booster* or *acoustic coupler*, is mounted between the electro–mechanical transducer and the tool. The vibrations amplitude at the output of the converter being about 1 nm to 100 nm [8], an amplification is necessary to obtain enough amplitude at the tool–tip workpiece interface. Therefore the sonotrode amplifies and transmits the vibrations from the transducer to the tool. Its geometry and dimensions are set to ensure the adjustment of its natural frequency to the generators excitation frequency.

For exponential flare of the sonotrode taper, the amplification is proportional to the ratio of the areas of the upper and lower faces of the sonotrode [9]. Machining tools or pin holding grippers can be screwed into this transducer. So a typical tool tip amplitude of 20 µm to 50 µm within a frequency range of 18 kHz to 70 kHz can be achieved.

4   Experimental Setup

The press–fitting, should it be conventional or ultrasonic, is characterized by its thrust force against feed curve as shown in **Fig. 1**. Therefore one needs to be able to drive the tool in the feeding direction and measure its displacement as well as measure the thrust force seen by the sample. For this reason the authors have designed, realized and mounted the semi-industrial ultrasonic press–fitting setup described here.

Vertical Movement. This setup consists of a *BOSCH REXROTH 3.842.993.178* rigid frame which carries a vertical linear guideway and two joined carriages. A *FANUC powerMotion i–A* numerically controlled NC axis drives a *PROMESS 64002–2201* ball screw to move these carriages. The nodal point of a 40 kHz *BRANSON GE101–135–67R* resonator unit is fixed to the carriages. The sonotrode is aligned on the same axis than the pin, the hole and the *KISTLER 9213B* force sensor. As shown in **Fig. 3**, this axis is also orthogonal to the samples upper face. The tool displacement is measured by a *HEIDENHAIN LS–487* linear encoder allowing an accuracy of less than 1 μm.

Force Measurement. Using a force sensor with a high cutoff frequency which is placed directly under the sample, it is possible to measure two components of the force view by the sample: the low frequency component due to the insertion movement and the high frequency component due to ultrasonic vibrations.

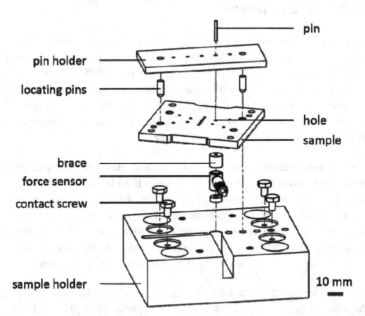

Fig. 3. The experimental setup includes a sample holder allowing a pre–positioning of the pins in the pin holder as well as a three point contact between the samples lower face and the sample holder using two of the four contact screws and the brace above the force sensor.

Holes. Due to the variability in hole diameters, and thus in interference, each experiment should be ran at least three times under the same conditions. Therefore the authors have bought sixty *B50* $CuZn_{39}Pb_2$ brass samples from *EMP* in Tramelan (Switzerland). Each of them contains 13 holes machined at EPFL – LPM with a special *Sphinx 51200* drilling–boring bit mounted on an industrial *Mikron HSM–400U* NC machine. The sample thickness is 1000 µm ± 10 µm. The nominal hole diameter is 990 µm ± 2 µm. Each hole has a 0.2 mm x 45° chamfer at both extremities to smooth the pin insertion into the hole.

Pins. To study the ultrasonic press–fitting at different interference values, three batches of pins have also been bought from *Adax*, in Bevaix (Switzerland). The pins are in 100Cr6 steel with a hardness of 60 HRC and a roughness of Ra 0.1. The nominal pin diameters are 997 µm ± 2 µm, 1000 µm ± 2 µm and 1020 µm ± 2 µm. Their length is 10 mm ± 100 µm with a 0.1 mm x 45° chamfer at each extremity.

5   Experimental Results

For this article the authors have executed the same experiment 16 times with success and good repeatability in the thrust force against feed curve shape. A typical result is shown in **Fig. 4**.

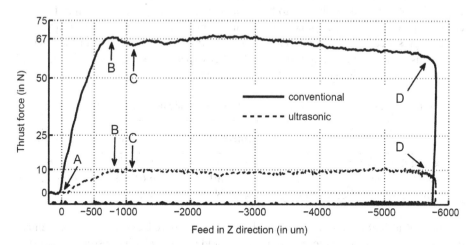

Fig. 4. The same steps as in conventional press–fitting are taking place: **A** Pin–hole chamfer contact **A to B** Feeding **B** Pin reaches minimal hole diameter **C** Pin reaches hole exit **C to D** Feeding **D** Pin reaches its final position **Comment.** The thrust force drops of 68 N to 11 N when ultrasound are used during press–fitting **Process parameters.** 16 runs, $CuZn_{39}Pb_2$ brass sample, sample thickness 1000 µm, mean hole diameter 991 µm, 100Cr6 steel pin, mean pin diameter 998 µm, pin roughness Ra 0.1, mean interference 7 µm, feed rate 10 mm/s.

Experimental Procedure. For both conventional and ultrasonic press–fitting, each run is executed with the same procedure described here.

1. The sample is placed on the sample holder and the pin is pre–positioned in contact with the sample. Locating pins on the pin holder guarantee alignment of the pin with both hole and force sensor as shown in **Fig. 3**.
2. Press–fitting operation runs according to the NC program: tool approaches up to 0.5 mm above the upper extremity of the pin and then moves down by 6 mm at a constant feed rate of 10 mm/s.
3. After reaching the lowest position, corresponding to **D** in **Fig. 4**, tool stops during 50 ms before moving off.
4. For runs carried out with ultrasound, the ultrasound generator is on during the whole press–fitting operation.

After each press–fitting, the maximal axial load before fastening failure is measured by turning over the sample and running the same NC program at a feed rate of 0.5 mm/s. The maximum axial load before sliding of the pin in the hole corresponds to the maximal axial load before the press–fit failure.

6   Discussion and Future Work

Discussion. As shown in **Fig. 4**, the typical thrust force against feed curve shape of a conventional press–fitting is also to be recognized in ultrasonic press–fitting. However, there is under the exact same experimental conditions a drastic drop in the thrust force in presence of ultrasound. This could be explained by a change in the friction conditions between the pin and the hole. Further experiments are needed in that field to confirm this hypothesis and build a reliable model.

Table 2. Comparison between conventional and ultrasonic press–fitting characteristics

	Conventional	Ultrasonic	Gain
Maximal thrust force	67.75 N	10.56 N	> 6x
Press–fitting energy	51.48 mJ	6.61 mJ	> 7x
Maximal axial load before sliding	64.91 N	46.94 N	≈ 0.7

Looking at **Table 2**, one can also observe a serious reduction in the mechanical energy needed to achieve the insertion. This could specially be interesting in industrial cases where the pin length to pin diameter ratio is high. Indeed, for such components a buckling is often observed. Reducing the thrust force and the press–fitting energy could overcome this buckling problem and extend the suitability of press–fitting for even smaller diameter pins.

Future Work. As the presented assembly technique is a novel one, there is still a certain amount of work to achieve in order to fully understand the influence of ultrasound on the press–fitting. The authors will continue to study this promising field of research and undertake the following actions:

1. Run a screening design of experiments to point out most influential process parameters which could be the sample material, hole depth, hole diameter, interference, feed direction, feed rate, vibrations amplitude, etc...
2. Run a second design of experiments to study the effect of the previously identified process parameters
3. Continue their study on the effect that manufacturing techniques have on the hole shape and hole diameter
4. Investigate for an alternative measurement system to gauges which would present a lower measuring instrument dispersion σ_m than 1 µm for (sub)millimetric holes

References

1. Astashev, V., Babitsky, V.: Ultrasonic Processes and Machines Dynamics, Control and Applications. Foundations of Engineering Mechanics. Springer, Berlin (1988)
2. Comprendre le chassage à l'échelle horlogère. In: Congrès International de Chronométrie, Montreux, Switzerland, Neuchâtel, Switzerland, Société Suisse de Chronométrie (October 2004)
3. Bourgeois, F.: Vers la maîtrise de la qualité des assemblages de précision. PhD thesis, Ecole Polytechnique Fédérale de Lausanne (2007)
4. Novel 3D analogue probe with a small sphere and low measurement force. In: ASPE Topical Meetings, Coordinate Measuring Machines, Charlotte, NC, USA. American Society for Precision Engineering, Raleigh (2003)
5. Montgomery, D.C.: Statistical Quality Control, 7th edn. Wiley, Oxford (2009)
6. Shaw, M.C.: Ultrasonic grinding. Microtechnic 10(6), 257–265 (1956)
7. Balamuth, L.: Ultrasonic rotary drive open up many new applications for micro-devices. Electronics 36(2), 56–57 (1963)
8. McGeough, J.A.: Advanced methods of machining. Chapman and Hall, London (1988)
9. Neppiras, E.A., Foskett, R.D.: Ultrasonic machining - technique and equipment. Philips Technical Review 18(11), 325–334 (1957)

Precision Micro Assembly of Optical Components on MID and PCB

Jonathan Seybold, Ulrich Kessler, Karl-Peter Fritz, and Heinz Kück

Hahn-Schickard-Gesellschaft, Institute for Micro Assembly Technology (HSG-IMAT),
Stuttgart, Germany
{seybold,kessler,fritz,kueck}@hsg-imat.de

Abstract. At HSG-IMAT the precision assembly of micro systems became increasingly important in recent years. The challenge is to reach a high accuracy of about 10 µm between the assembled elements. At the same time it is important that the applied processes are suitable for low cost manufacturing. Attempts were made to achieve these requirements by means of automatic assembly processes based on MID and PCB. The results of the development at HSG-IMAT are shown in this contribution using the example of an optical module of a rotary encoder. Thereby, the assembly of the laser diode and lens requires special attention to achieve the specified positioning tolerances.

Keywords: Optical module, laser diode, VCSEL, sensor, rotary encoder, precision assembly, low cost, miniaturization, MID.

1   Indroduction

Micro assembly of optical elements often requires high precision joining operations during the assembly of the components. In this contribution the assembly of the optical components of a module for rotary encoders is described. Rotary encoders are used in many industrial applications for positioning and motion control. Thereby, optical encoders are usually used for high precision measurements [1].

In this case, an optical encoder was designed and built up with an alternative functional principle, which was newly developed at HSG-IMAT. The first feature is the precisely micro structured plastic disc as solid measure similar to a compact disc (CD) or digital versatile disc (DVD). The disc of the encoder is made of polycarbonate and is manufactured by injection moulding at low manufacturing costs.

The second feature is that the encoder contains an optical module with a laser diode, a lens and photo diodes. The assembly of this optical module is described in this paper. The motivation of the research is to reach a plug and play assembly of the encoder itself which leads to a low cost encoder but in spite of this, the encoder has also a high resolution [2]. To achieve this objective, a relative positioning accuracy of ±10 µm between laser diode and lens is required in the optical module. The reason for this is, that all critical tolerances than are included in the optical module itself.

To achieve the required accuracy of ±10 µm between laser diode and lens two different approaches were proposed. On the one hand an encoder with a MID optical

S. Ratchev (Ed.): IPAS 2014, IFIP AICT 435, pp. 30–36, 2014.

module was designed and built up [3]. In this case, the challenge is to reach the required accuracy during the assembly process of the laser diode. On the other hand the
optical components were directly assembled on PCB with automated assembly
processes [4]. In this case, the challenge is to reach the required accuracy during the
assembly process of the lens.

2   Optical Module on MID

The first optical module was based on Moulded-Interconnect-Device (MID) technology using the LPKF LDS® process [5]. The optical module contains one laser diode
(VCSEL), one lens and five photo diodes. First, the laser diode was manually attached
exactly to the center of the MID-header within the required tolerance of ±10 µm using
a Finetech Fineplacer. Accordingly, the five photodiodes were manually attached
whereby the cathode was connected with conductive adhesive. All dies were wire-
bonded directly to the MID-substrate to connect the anode. Afterwards, the lens was
assembled manually with form closure on the MID-header. In the final step, the optical module was connected by conductive adhesive to a PCB on which the electrical
processing of the signals was performed. Fig. 1 shows on the left side the assembled
and wire-bonded dies on the MID-substrate and below the assembled lens. The right
side shows the completely assembled optical module together with the PCB. The
photo diodes with an edge length of 1.4 mm can be used as scale.

The contour of the lens has been designed to fit on the MID-header at the bottom
and to carry an aperture at the top. The lens was made of PMMA and was manufactured by ultra-precise diamond turning. For larger quantities the lens can also be manufactured by injection moulding.

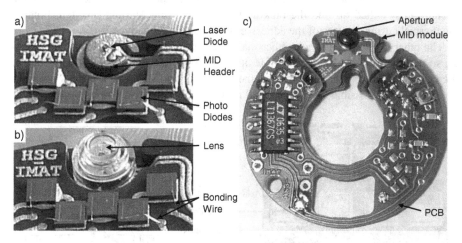

Fig. 1. a) MID-substrate with assembled laser diode and photo diodes; b) MID-substrate with
assembled lens; c) MID module inclusive the assembled aperture connected to the PCB

3   Optical Module on PCB

3.1   Assembly of the Laser Diode

Since the first results of the output signals of this encoder were very promising, further investigations on this encoder technology including the precise assembly technology were started at HSG-IMAT. The next step was to apply the optical elements on PCB. Therefore the assembly of the laser diode has been studied in detail. Because of its small size of only 200 x 200 x 150 µm³ the automated assembly of the laser diode is a challenge. The investigations were conducted on the automated micro assembly system Vico Base from Häcker Automation GmbH.

At first, different vacuum pick up tools for the assembly of the laser diode were tested. The best results were achieved with a pick-up tool with conical shape which holds the laser diode at its edges. Thereby the laser diode is held in a defined position in the tool through a self centering effect because of the conical shape of the tool [6]. This allows to assemble the laser diode without a bottom side referencing step. It has become evident, that a bottom side referencing of the laser diode leads to a loss of positioning accuracy due to the poor quality of the bottom side edges in combination with an insufficient resolution of the inspection system. Fig. 2 shows on the left side the laser diodes on blue tape in the state as delivered. The right side schematically shows the pick-up process of the laser diodes from the blue tape.

The assembly process of the laser diode runs fully automatic in several steps. First the reference marks on the PCB are detected with the inspection camera. Subsequently the conductive adhesive is applied to the target position by a stamping process.

Afterwards the respective pick-up position of the laser diode on blue tape is detected exactly with the inspection camera and the laser diode is picked up and held in the pick-up tool by vacuum. After moving to the target position and the assembly the vacuum is switched off and the pick-up tool moves to the starting position. Finally, the thermal curing process of the adhesive is carried out in an oven.

The other studied tools were either made from rubber or metal and had flat surfaces, so that the tool touches the top side of the chip. With the metal tool the surface of the laser diode was damaged and with the rubber tool no stable pick-up and positioning process was possible due to the small rubber tip, which proved to be too instable.

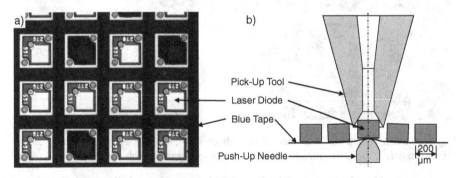

Fig. 2. a) Laser diodes on blue tape; b) Schematic pick-up process from blue tape

3.2   Positioning Accuracy of the Laser Diode

To investigate and optimize the positioning accuracy several 3x3 laser diode arrays were assembled on an experimental PCB using different setups. In detail the shape and size of the reference marks on the PCB were varied and also the procedure of the camera inspection of the laser diode before the pick-up process from blue tape. Fig. 3 shows such an assembled laser diode array.

It could be shown, that the automatic inspection works best with rectangular or linear contrast transitions. Therefore rectangular reference marks should be preferred over circular or crosswise marks. Accordingly the inspection of the laser diode works better when the rectangular structures of the surface are used instead of the circular alignment marks. Furthermore it has been shown that for highly precise assembly it can be advantageous to use the reference marks on the PCB only for assembly of the first die if there is more than one chip to be assembled. For the subsequent assembly steps it is then possible to use the first assembled die as reference mark. In this way, the more precise lithographic structures of the die instead of the etched reference marks on the PCB, which are less precise, are used to achieve the required accuracy.

The 3x3 laser diode array in Fig. 3 was assembled according to this principle. The first laser diode was assembled in relation to the reference mark on PCB. The other laser diodes were assembled in relation to the first laser diode, whereby for each laser diode a new inspection step of the reference laser diode was done. In order to investigate the quality of the optical referencing, the PCB was loaded to and unloaded from the machine between the assembly of each laser diode.

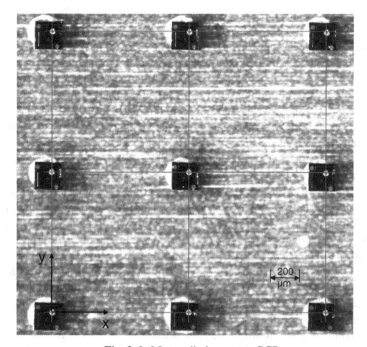

Fig. 3. 3x3 Laser diode array on PCB

The positions of the facets of the laser diodes were measured before and after the curing process on a coordinate measuring machine und compared with the intended target positions. The diagram in Fig. 4 shows the measured displacement of the facets of the laser diodes from their target positions. After curing, the maximum displacement was found to be ±7 µm, the standard deviation was < 3.5 µm. Altogether five 3x3 laser diode arrays were assembled.

The die attach of the photo diodes is easier due to their larger size. In addition, the sensor concept doesn't require the same level of accuracy as it is with the laser diode and lens. The assembly could therefore be done using a rubber tool.

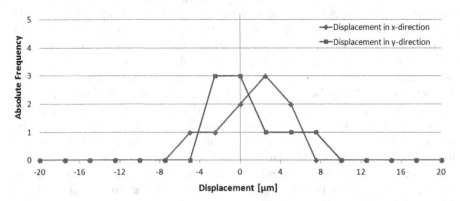

Fig. 4. Frequency distribution of the displacement in x- and y-direction

3.3   Assembly of the Lens

The assembly of the lens has similar requirements concerning the accuracy like the laser diode. First, UV curable adhesive is transferred by a ring-shaped stamp to the PCB. Then the lens is picked up with a conical pick-up tool which assures the defined alignment of the lens in the pick-up tool. The target position of the lens on the PCB equates to the facet of the laser diode. Therefore an inspection step of the laser diode is done. Then the lens is assembled and the adhesive is cured through UV radiation. To ensure the correct distance between laser diode and lens, on the bottom side of the lens a gap for the adhesive is defined. Fig. 5 shows schematically the assembly process of the lens on the PCB.

The challenge in this case is to find the center of the lens before the pick-up step. Because of the clear optical surfaces, it is difficult for the inspection camera to find sufficient contrast at the edges. For this case, a ring at the top side of the lens was designed. When manufacturing the lens by injection moulding, the ring shaped plane can be arranged with a mat surface to enhance the contrast.

The assembled optical module on PCB is shown in Fig. 6. Visible are the PCB with the electronic components, which were assembled first, the photo diodes and the lens. The laser diode is located below the lens. The photo diodes with an edge length of 1.4 mm can be used as scale.

The misalignment between laser diode and the assembled lens is determined as the difference between the center of the lens, measured by optical inspection and the centre

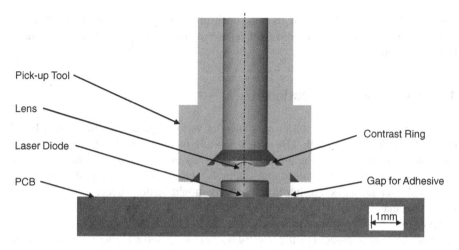

Fig. 5. Schematic assembly process of the lens on PCB

Fig. 6. Complete assembled optic module on PCB

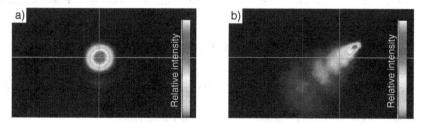

Fig. 7. Beam profile of a correct assembled lens *(a)* and of a misaligned lens *(b)*

of the facet of the laser diode which can be seen as image through the lens in its focal plane. Another method is to consider the beam profile of the spot. Fig. 7 shows on the left side the beam profile with a correctly aligned lens and on the right side a misaligned lens. Shown is the relative intensity of the laser beam in its beams waist plane in which is the encoder disc with its solid measure too. A misaligned lens causes of course poor signal quality or even total failure of the rotary encoder when the misalignment is too large.

4   Conclusion and Outlook

A miniature optical module for a rotary encoder was designed and built up first based on MID and further on PCB. The optical module includes one laser diode, one lens and several photo diodes. The positioning accuracy of the assembled optical components was analysed. The result is that assembly within ±10 µm is possible under the described circumstances. Further investigations should be done under the condition of large scale manufacturing to confirm this result.

HSG-IMAT will continue to work in the field of precision assembly. For example automated active alignment of the lens by beam control and a higher resolution of the inspection camera of the 3-D assembly machine are two issues, which will bring further progress in precision assembly.

In addition, it is planned at HSG-IMAT to investigate and develop forward the 3-D assembly on MID. This enables to use the advantage of the MID-technology in reference to the degree of freedom in 3-D. From this point of view, this technology allows to design mechanical joining elements which can be used for the purpose of alignment or also actuators and sensors with functions based on 3-D, for example 3-D magnetic field sensors.

References

1. Schaeffler, K.G.: Magnetisch versus optisch; Winkel-Mess-Systeme für Rundtischachsen im Vergleich (2008), http://www.schaeffler.com
2. Hopp, D.: Inkrementale und absolute Kodierung von Positionssignalen diffraktiver optischer Drehgeber; Dissertation Universität Stuttgart (2012)
3. AiF-Vorhaben Nr. 219 ZN: Untersuchungen zu einem hochauflösenden optischen Drehwinkelsensor in Low-Cost-Bauweise; Institut für Mikroaufbautechnik; Hahn-Schickard-Gesellschaft (2008)
4. AiF-Vorhaben Nr. 349 ZN: Untersuchungen zu kostengünstigen absolut und inkremental kodierten optischen Drehgebern mit justagefreier Endmontage und mikrostrukturierter diffraktiver Maßverkörperung aus Kunststoff; Institut für Mikroaufbautechnik; Hahn-Schickard-Gesellschaft (2012)
5. Forschungsvereinigung 3-D MID e.V.: 3D-MID Technologie. München, Carl Hanser Verlag (2004)
6. Die Attachment and Fluid Dispensing; Brochure; SPT Roth Ltd. (2013), http://www.smallprecisiontools.com

Integrated Tool-Chain Concept
for Automated Micro-optics Assembly

Sebastian Haag[*], Tobias Müller, Christoph Pallasch, and Christian Brecher

Fraunhofer Institute for Production Technology IPT, Steinbachstr. 17,
52074 Aachen, Germany
sebastian.haag@ipt.fraunhofer.de

Abstract. The work presented in this paper aims for a significant reduction of process development and production ramp up times for the automated assembly of micro-optical modules and systems. The approach proposed bridges the gap between the product development phase and the realization of automation control through integration of established software tools such as optics simulation and CAD modeling as well as through introduction of novel software tools and methods to efficiently describe active alignment strategies. The focus of the paper is put on the formalized specification of product configurations as a basis for the engineering of automated assembly processes. The concepts are applied to industrial use-cases. The paper concludes with an overview of the application of the concepts in an engineering tool-chain as well as an outlook on the next development steps.

1   Introduction

Photonics technologies are key technologies of the 21st century enabling innovative applications in demanding and societal relevant domains. In many cases, photonic applications belong to the field of high-technology applications with small and medium production volumes. In order to respond to increasing competitive pressure in production of micro-optical systems and lasers, flexible, autonomous, and efficient solutions for assembly systems have to be developed [1]. Competitive small and medium size volume production can only be achieved based on an increased degree of automation as automation can reduce costs while realizing highly reliable processes and high product quality as well as improved working conditions. However, a high degree of automation still correlates negatively with the flexibility of a production system which can lead to a major negative impact on the return on investment calculation. Recently, promising progress has been made developing modular and reconfigurable assembly systems [9, 11]. There is a significant lack of corresponding software solutions to allow for an efficient exploitation of the available flexibility as promised by computer-aided process planning (CAPP).

In order to fulfil precision requirements and due to manufacturing tolerances and finite accuracy of the positioning systems, the assembly of micro-optical systems

---

[*] Corresponding author.

S. Ratchev (Ed.): IPAS 2014, IFIP AICT 435, pp. 37–46, 2014.

requires the use of sensor-guided processes such as vision-based component pickup, passive alignment exploiting geometrical features through image-processing, and active alignment optimizing the optical function towards a desired value. The use of sensors requires the processing and analysis of signals and in some cases reasoning in order to extract information from it which drastically increases the complexity of process planning and commissioning. Thus, the required engineering time is high and nullifies the benefit of automation. This is the motivation for ongoing research activities which intends to develop and validate an integrated development environment (IDE) for automated precision assembly processes bridging the gaps between computer-aided product development, virtual process planning and assembly execution by means of methods and software tools for implementing complex sensor-based handling and alignment processes.

The assembly of micro-optical systems is still dominated by manual operations. Feasibility of automation has been proven for many cases such as resonator mirror alignment, focus optics alignment, fast-axis collimator alignment for diode laser bars as well as for multi-single emitter diode laser modules ([2, 3, 4, 5, 6]). Investment and implementation costs are still high and mean a high risk – especially for small and medium sized companies. An engineering gap regarding the information transfer from product development to the subsequent engineering phases of process development and assembly system configuration and commissioning can be stated. This gap needs to be overcome in order to allow automation technology to achieve a breakthrough in the production of optical systems such as lasers. The work presented in this paper strives for closing the identified engineering gap by proposing a standardized and yet flexible approach for engineering such assembly solutions supported by corresponding software tools which do not exist in the domain of laser system production. The main task identified by the authors is the introduction of a flexible engineering methodology for planning and commissioning the automated assembly of optical systems and therefore the integration of existing tools and the conceptual design and the prototypal implementation of novel tools for bridging existing engineering gaps.

2   Concept of Hybrid Assembly Process Planning for Optical Systems

This paper proposes a hybrid approach for assembly process planning for optical systems in which parts of the engineering can be automated as in CAPP and other engineering tasks are implemented or optimized by operators. ElMaraghy defines in [11] the term of micro-process planning as the specification of detailed parameters for individual operations. This definition corresponds with the intent of the proposed approach which integrates well-established engineering tools (for CAD modeling, optical simulation, PLC and robot programming) as well as novel software modules for the specification of sensor-guided process parameters through standardized communication interfaces and data formats. An iterative approach of virtual and physical engineering steps is proposed in order to provide sufficient flexibility during the process planning phase. Virtual engineering steps comprise offline and computer-aided development efforts. Physical engineering steps stand for development efforts carried out physically, e.g. the refinement of an active alignment process using real laser and camera (instead of simulation). The term 'online' is avoided because the authors use that term for activities in the actual production system.

Product Design (virtual) PD	Product Prototyping (physical) PP	Virtual Platform-independent Process Planning vPIPP	Physical Platform-independent Process Planning pPIPP	Virtual Platform-specific Process Commissioning vPSPC	Physical Platform-specific Process Commissioning pPSPC
- *Design of optical system* - *CAD including thermal analysis*	- *Assembly and evaluation of prototypes* - *Design review*	- *T2B KC-specification* - *Assembly sequence generation* - *Alignment-in-the-loop*	- *Realization of alignment strategy* - *Enhancement of alignment towards robustness*	- *Teaching in VR* - *Allocation of represented hardware entities*	- *Integration of hard- and software*

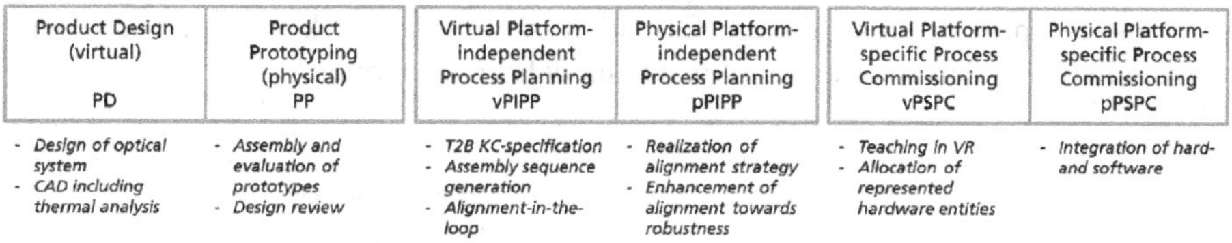

Fig. 1. Overview of engineering elements for hybrid planning and commissioning the automated assembly of optical systems

Fig. 1 provides an overview of the elements of the proposed engineering platform. Product design (PD) is usually carried out using computer-aided (virtual) methods. Early physical realization of the product as a prototype allows evaluation of the design and possibly design improvements. For process planning and commissioning the authors propose a two-stage approach analogue to model-driven engineering approaches for maximizing re-usability [12]. The first stage is a platform-independent view (xPIxx) on the task while the second stage is platform-specific (xPSxx). Furthermore, process planning and commissioning are both proposed to be subdivided in a virtual as well as a physical branch. This framework leads to four engineering elements:

- Virtual platform-independent process planning (vPIPP): This engineering step carries out top-to-bottom (T2B) specification of key characteristics in a formal way and uses that information for deriving an assembly sequence. For the development and verification of specific assembly steps, tools such as the proposed alignment-in-the-loop can be used to virtually specify and verify complex tasks offline.
- Physical platform-independent process planning (pPIPP): On a dedicated prototyping platform individual process steps such as active alignment pre-planned in the virtual environment can be executed and validated. Discrepancies between ideal models and the real world can be identified and compensated this way. A possible solution to overcome such discrepancies is the integration of visible reference marks for local referencing between model and hardware.
- Virtual platform-specific process commissioning (vPSPC): For mapping the assembly task specification consisting of an assembly sequence and virtually planned process steps to an assembly platform, the concept proposes the use of Virtual Production Systems (VPS). VPS's can be used for teaching coordinates in the model environment. High-precision tasks require the use of sensor-guidance based on visible reference marks.
- Physical platform-specific process commissioning (pPSPC): The final step of commissioning takes place on the production platform in a drastically reduced way complexity and ramp up time.

Fig. 2 depicts the engineering phases of product development, process planning, and commissioning of the assembly task in a slightly different way than in the previous section. Here, the paths of information exchange are illustrated by arrows. For an efficient engineering process, it must be ensured that all phases have well-defined interfaces for passing on and exchanging information between each other. Note, that no direct links for information exchange are depicted between the physical engineering steps (PP, pPIPP and pPSPC). Information must be fed back to the virtual environment in the models in order to refine the model-driven approach.

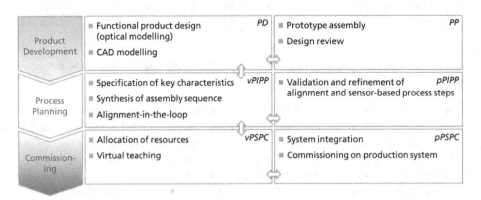

Fig. 2. Engineering phases for the assembly of micro-optical systems. Arrows indicate flow of information.

2.1 Overview of the Integrated Tool-Chain

For the development of the photonic products, established software tools are available such as optical simulation tools and CAD modeling tools (Fig. 3, left column). Assembly process planning for micro-optical systems comprises the sequence of assembly steps and the planning of processes such as part feeding and alignment. The product configurator edits formalized representation of the liaison diagram and therefore bridges the gap between product development and process planning (Fig. 3, middle column). In the VPS resources are assigned and the assembly sequence is verified regarding logic and collisions (Fig. 3, right column).

PD	vPIPP	vPSPC
Optical modeling and simulation	Product configurator (specification of KC's)	Virtual Production System
Mechanical design and thermal analysis	Alignment-in-the-loop	
	Toolkit for detection of reference marks	

Fig. 3. Overview of tools in IDE

2.2   Formalization of the Product Configuration

Product development consists of several phases. In a first step, functional and non-functional requirements are defined from a customer point of view. In a second step, the defined requirements are transformed into technical requirements specifications. In a third step, for optical systems, an optical model is created and assembly tolerances are analyzed through simulation. As a result of this step the optical elements are chosen and the product can be designed in CAD regarding its outer and inner dimensions. Furthermore, the mechanical design of laser systems is optimized regarding its thermal behavior by means of thermal analysis. Professional and well-established software is available for all of these steps.

In the case of optical systems, special attention needs to be paid to the high precision demands often reaching in the sub-micrometer domain. Process steps such as handling, alignment and bonding of parts in many cases need sensor-based strategies compensating for inaccuracies induced by unavoidable mechanical tolerances. Therefore, in case of optical systems, the authors adapted ideas of classical process planning in an intermediate step describing the desired geometric and especially the desired optical relations of the product parts for bridging the gap between product development and assembly execution. According to the terminology defined in [7] for mechanical assemblies, constraints or requirements on geometric and mechanical relations are named *key characteristics* (KC's).

KC's express the objectives and constraints of individual assembly steps as well as a quality measure for the whole assembly. This paper proposes to additionally include optical functions in the set of key characteristics. One important difference compared to plain mechanical assemblies is that constraints can be defined by parts that do not have a direct physical linkage. The modeling environment needs to support these circumstances. Furthermore, this paper shows how to formally define dependencies between KC's allow the derivation of the sequence of assembly steps (assembly sequence graph or precedence graph) and therefore the generation of platform independent assembly logic. KC's defining the desired relation between certain parts can be modeled hierarchically in order to derive mechanical constraints from optical ones. This information completes the assembly logic.

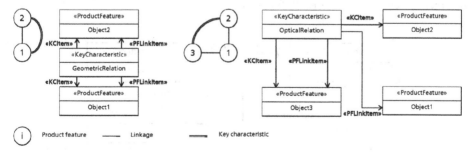

Fig. 4. Formalized target configuration patterns as liaison diagram and as formalized liaison diagram: key characteristic directly specifying the constraints on a physical linkage between two product features (left), key characteristic specifying the constraints on a physical linkage between two product features using a third product feature (right).

Fig. 4 shows two patterns of describing physical linkages between product features such as gluing areas or optical surfaces. Two types of representations are used in the figure. One representation (liaison diagram) uses a number code representing a product feature. The solid line connecting two product features stands for a physical linkage, the double line stands for a key characteristic specifying the constraints of the physical linkage. A constraint could be a specific geometric relation or an optical function. This kind of representation is applicable for intuitively sketching out target configurations.

The second representation introduced by the authors is a formal representation of a liaison diagram (formalized liaison diagram) and it uses a newly defined UML profile and is semantically equivalent to the first one. Extending UML allows the use of standard computer-aided software engineering (CASE) tools. State-of-the-Art UML tools allow the application of text templates on such diagrams for generating machine readable code which can be processed further within the subsequent engineering tool-chain.

By definition in this paper, one KC physically links two product features (PFLinkItem). Furthermore, a KC defines a geometric or a functional relationship. In order to describe such a relationship two more product features are referenced as key characteristic items (KCItem). KCItems and PFLinkItems can be the same product features as depicted in Fig. 4, left. This is the case when, for example, an optical element is bonded onto a carrier plate using geometric features such as parallelism and distance of edges. Sometimes it is necessary to physically link two product features while the key characteristic such as an optical function is defined by other items. Fig. 4, right, shows a configuration where Object1 (carrier) holds Object2 (lens) and Object2 is attached to Object3 (e.g. heat sink of a laser diode bar). The constraint on the physical linkage between Object1 and Object3 is determined by the optical function between Object3 and Object2.

The alignment of micro-optical components fulfilling optical constraints often requires multiple steps. Firstly, the positioning system places the optical component using machine accuracy (pre-positioning). Secondly, the component is aligned according to geometric features (passive alignment) using machine vision and e.g. edge detection. Thirdly, many optical components require active alignment for directly optimizing the desired optical function. Fig. 5 shows an optical KC which is refined by other key characteristics (KCRefinement).

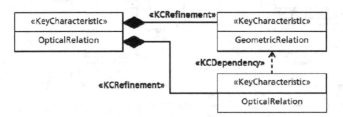

Fig. 5. Refinement of key characteristics (KCRefinement) and dependencies between key characteristics

Using the proposed UML profile, dependencies between key characteristics can be expressed. This is required for describing multi-step alignment (pre-positioning, passive alignment, active alignment). The combination of the features of refinement and dependency allows the expression of highly complex assembly constraints carrying all relevant information for subsequent assembly execution. Based on the formalized product configuration description the assembly sequence can be derived as explained in the following section.

2.3   Derivation of the Assembly Sequence and Commissioning

The assembly sequence can be derived from the dependencies between key characteristics by evaluating the KCRefinement and KCDependency relation stereotypes in the formalized liaison diagram. The resulting assembly sequence describes the assembly task in a platform-independent way because the assembly platform is not part of this description (vPIPP) and therefore no platform specific information is available. Neither resources such as manipulators and grippers nor coordinates are specified yet. Active alignment routines and other complex process steps are represented by placeholders at first. Within the context of the work presented, different tools are under development for efficiently planning sensor-based processes in a hybrid assembly process planning system.

Once an alignment strategy is referenced as a subroutine by the placeholder its functionality is verified by an alignment-in-the-loop software tool. This tool uses the application programming interface of the commercial optical modelling and simulation software for manipulating the position and orientation of the lens under inspection. Output of the simulation is a simulated intensity profile of the optical setup which is processed by the alignment routine calculating corrective values for manipulating the lens. For the verification of alignment strategies, the lens under inspection is initialized randomly regarding its position and orientation and then the steps of processing the intensity profile, calculating the corrective value, and manipulating the lens virtually are carried out in a loop until the termination criteria is reached.

The transfer to a production platform requires two major steps: 1. teaching of coordinates in the VPS, 2. assigning reference mark detection routines. Teaching and planning based on a VPS minimizes standstill times of the physical production system. **Fig.** 6 shows different shapes of reference marks which can be detected in a standard way. This allows a mapping between model data and real world data during process execution.

Fig. 6. Samples of reference marks with corresponding results from image-processing (from [4])

3   Exemplary Assembly Specification

This section illustrates the application of the proposed formalism for specifying the assembly task of a collimation module for diode laser systems. The optical system consists of a fast-axis collimation lens (FAC) and a slow-axis collimation lens (SAC). The optics have to be aligned in a specific distance in order to fulfill the optical function of the module. Two side tabs are used for holding the optical components. The side tabs do not have an optical function. Fig. 7 shows the target configuration of the collimation module as a wire frame model (middle) and its representation in a liaison diagram (left).

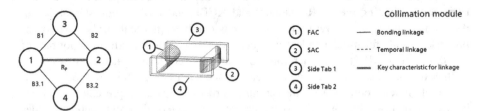

Fig. 7. Representation of the target configuration of a collimation module. The wire frame sketch (middle) illustrates the optical system. The left column shows the formal representation.

Fig. 8 depicts the assembly sequence of the collimation module. The additional nodes (A) and (B) represent mechanical end stops used for passive referencing. The dashed lines represent temporal linkages used during assembly. In step (a), the side tab is aligned temporarily using the mechanical endstops (A) and (B). In step (b), the FAC is inserted temporarily. In (c), the linkage is controlled by a key characteristic applying geometric constraint. In (d), the SAC is assembled regarding a geometric constraint between FAC and SAC. Steps (e) and (f) add the second side tab to the module. Picture (g) represents the final product with the fulfilled constraint.

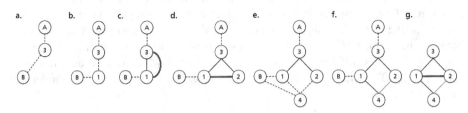

Fig. 8. Individual assembly steps of the collimation module with side tabs

4   Conclusion and Outlook

This paper introduced a formal notation for specifying target configurations of micro-optical systems. The notation includes information about key characteristics which determine the optical function of the system as well as the constraints of each assembly step. Key characteristics can be refined in order to match the multistep alignment approach for optical systems comprising the steps of pre-positioning, passive alignment, and active alignment. Furthermore, the definition of dependencies between key characteristics can be exploited in order to generate a platform-independent assembly sequence.

Future work will be dedicated to the realization of a tool-chain by implementing software tools for specifying the individual assembly steps in a level of detail which allows their execution through an assembly platform. The focus will be put on the specification of strategies for sensor-guided process steps and especially active alignment processes as the engineering of such processes is currently the bottleneck regarding the planning and commissioning of micro-optical assembly.

Acknowledgement. The authors would like to thank the German Federal Ministry of Education and Research (BMBF) for the support of the work by financing the MA-NUNET-DeLas project (funding code 02PJ2542).

References

[1] Brecher, C. (ed.): Integrative production technology for high-wage countries. Springer, Berlin (2012)

[2] Brecher, C., Pyschny, N., Haag, S., Guerrero Lule, V.: Automated alignment of optical components for high-power diode lasers. In: Zediker, M.S. (ed.) High-power diode laser technology and applications X, San Francisco, California, United States, January 22-24. Proceedings of SPIE, vol. 8241, pp. 82410D–82410D-11. SPIE, Bellingham (2012)

[3] Brecher, C., Pyschny, N., Haag, S., Mueller, T.: Automated assembly of VECSEL components. In: Hastie, J.E. (ed.) Vertical external cavity surface emitting lasers (VECSELs) III, San Francisco, California, United States, February 3-5. Proceedings of SPIE, vol. 8606, pp. 86060I–86060I-14 (2013)

[4] Garlich, T., Guerrero, V., Hoppen, M., Müller, T., Pont, P., Pyschny, N., et al.: SCA-LAB. Scalable Automation for Emerging Lab Production. Final report of the MNT-ERA.net research project. Apprimus-Verl (2012)

[5] Miesner, J., Timmermann, A., Meinschien, J., Neumann, B., Wright, S., Tekin, T., et al.: Automated Assembly of fast-axis collimation (FAC) lenses for diode laser bar modules. In: Zediker, M.S. (ed.) High-power diode laser technology and applications VII, San Jose, California, United States, January 26-27, vol. 7198, pp. 71980G–71980G-11. SPIE, Bellingham (2009)

[6] Pierer, J., Lützelschwab, M., Grossmann, S., Spinola Durante, G., Bosshard, C., Valk, B., et al.: In: Zediker, M.S. (ed.) High-power diode laser technology and applications IX, San Francisco, California, United States, January 23-25, vol. 7918, pp. 79180I–79180I-8. SPIE, Bellingham (2011)

[7] Whitney, D.E.: Mechanical assemblies. Their design, manufacture, and role in product development. Oxford University Press, New York (2004)

[8] Scheller, T.: Analysis of Automated Assembly Processes for Hybrid Micro-optical Systems (German original: Untersuchung zu automatisierten Montageprozessen hybrider mikrooptischer Systeme). Dissertation (2001)

[9] Müller, R., Riedel, M., Vette, M., Corves, B., Esser, M., Hüsing, M.: Reconfigurable Self-optimising Handling System. In: Ratchev, S. (ed.) Precision Assembly Technologies and Systems, pp. 255–262 (2010)

[10] Gottipolu, R.B., Gosh, K.: Representation and Selection of Assembly Sequences in Computer-aided Assembly Process Planning. International Journal of Production Research 35(12), 3447–3466 (1997)

[11] ElMaraghy, H.A. (ed.): Changeable and Reconfigurable Manufacturing Systems. London: Springer London (Springer series in advanced manufacturing) (2009)

[12] Gašević, D., Djurić, D., Devedzic, V.: Model driven engineering and ontology development, 2nd edn. Springer, Dordrecht (2009)

Feeding of Small Components
Using the Surface Tension of Fluids

Matthias Burgard, Nabih Othman, Uwe Mai, Dirk Schlenker, and Alexander Verl

Fraunhofer Institute for Manufacturing Engineering and Automation IPA,
Stuttgart, Germany
{matthias.burgard,nabih.othman,uwe.mai}@ipa.fraunhofer.de

Abstract. The feeding of components smaller than $1\,\text{mm}^2$ is a high challenge in an automated manufacturing line. The surface forces affecting these micro components are getting more important and cause problems when handled. A new method is described using the surface forces for separating, sorting and arranging micro components. The surface shape of fluids and the gravity are used for moving the floating components to defined positions. The components can be arranged in a magazine or can be sorted by a channel system.

Keywords: micro, sorting, feeding, arranging, handling, fluid, surface force, surface tension.

1   Introduction

Due to the ongoing miniaturization of products and the simultaneously increasing scale of integration, the components to be handled are getting smaller. In contrast, the micro components and the assembly processes are continuously getting more sensitive towards the environment. As a result of this trend, the common vibratory conveyor technology reaches its limits due to its impact to the components and the occurring surface effects. In these cases, the separation and the arranging in a defined order for a further manual or automatic processing, for instance gripping of the component in the assembly process, is limited or even impossible.

In [1]-[3] new processes are described using surface tension for the assembly of micro parts. It has the big advantage that the surface forces are controlled and positively used for the process. In contrast the feeding of the micro components is still a conventional process with all its disadvantages for the handling of the small parts.

2   Approach

The fluidic sorting technology described here is an approach adapted for components smaller than 1mm which are provided as bulk good as for instance gear wheels, optical components, coated o-rings or electronic components (see figure 1). The fluidic sorting is based on the phenomenon to be found in nature that water striders can stay and glide on the water surface due to the surface tension. If small and therefore

S. Ratchev (Ed.): IPAS 2014, IFIP AICT 435, pp. 47–51, 2014.
© IFIP International Federation for Information Processing 2014

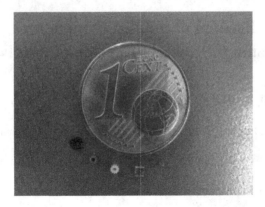

Fig. 1. Tested miniaturized components

lightweight components are applied on a surface of a fluid, it shows similar behavior. In case of a concave or convex curve, the components are sliding to lower level to the fluid rim or to a present barrier.

Fig. 2. Sliding o-rings on a convex fluidic surface

3   Functionality and Set-up

A defined formed fluidic surface is required for the process. To provide this, the fluidic sorting device features a reservoir which is incorporated into a base plate (see figure 3). The reservoir is filled with fluid and the required surface curvature is set by the filling level of the reservoir. After the reservoir is filled, the components will be applied onto the fluid. Due to the gravity, the components are moving to the rim of the fluid, which is defined by the geometry of the encircled incorporated structure.

An agglomeration of components can be split up by a defined vibration. As soon as all components are arranged, the level of the fluid is lowered. A barrier is included in the reservoir rim, which can be overflown by the fluid but prevents the components to float back with the sucked fluid. In this manner an arrangement is achieved to enable an extraction of the components for instance with tweezers or with a pick-and-place system.

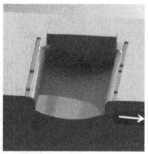

Fig. 3. Principle of the fluidic sorting process

The set-up consists of four significant components:

1. the fluidic sorting device,
2. the fluid supply unit,
3. the collector module and
4. a vibration unit.

Fig. 4. Set-up with fluidic sorting device, collector module and interfaces to the fluid supply unit

The fluidic sorting device contains the reservoir. With the connected pump and valve system of the fluid supply unit, the reservoir can be filled and drained. An essential element of the set-up is the collector module for the capture of the processed components. It can be inserted into the fluidic sorting device. The width and depth of the incorporated structure is adapted to the components and it can have additional cavities for the separated arranging of the components. In the case of a further external processing, the collector module can be designed as a magazine. The collector module can be closed with an adapted cover element and airflow can be passed through the channel, realized by closing the cover, to dry the components. Also alcohol or ultrapure water can be used for cleaning or to reduce the risk of residues or particle contamination. The whole set-up is located on a vibration unit. With defined frequency, it is possible to split up an agglomeration and to arrange the components in a row.

Fig. 5. Arranged components in a row after vibration

Components that where positively tested already are shown in figure 1. Not only the shape and dimensions of the micro components are manifold, also the material. Brass, glass, sapphire, silicone and various polymers (e.g. silicone) can be processed.

4   Conclusion and Outlook

A new process for separating, arranging and feeding micro components has been described using the surface tension for a gentle handling. A parallel processing of multiple parts is feasible with this method. Micro components made of various materials and many shapes can been processed with this method. A test platform has been realized including set-up and the control. In future tests, the process will be characterized in respect to e.g. reliability, failure ratio or throughput.

References

1. Burgard, M., Schläfli, N., Mai, U.: Processes for the self-assembly of micro parts. In: Ratchev, S. (ed.) Precision Assembly Technologies and Systems. IFIP AICT, vol. 371, pp. 36–41. Springer, Heidelberg (2012)
2. Burgard, M., Mai, U., Verl, A.: Automatisierte Bestückung von LEDs basierend auf Self-Assembly. In: Bundesminister für Bildung und Forschung: MikroSystemTechnik Kongress 2011, October 10-12. Darmstadt, pp. 559–561. VDE Verlag, Berlin (2011), nbn:de:0011-n-1834644

3. Burgard, M.: Self-Assembly - ein neuartiger Ansatz für die automatisierte Mikromontage. In: Schlenker, D. (ed.) (Tagungsleitung); Zeitschrift Mikroproduktion: mikroMontage: Prozesstechnik, Anlagentechnik. Fachtagung, Stuttgart, München, May 10-11. 22 Folien, Hanser, Wien (2011), nbn:de:0011-n-1850017

4. Schlenker, D., Othman, N.: Vereinzeln und Zuführen mikrotechnischer Bauteile. Mikroproduktion 2, 10-13 (2013)

5. Burgard, M., Othman, N., Frei, M.: Neue Herausforderungen an die Montagetechnik. In: MicroMountains: 5, February 27. iNNOVATION fORUM für Mikrotechnik, 11 Folien. Villingen-Schwenningen (2013)

6. Othman, N.: IPA.FluidSort - Separation of smallest componentsIn: MicroMountains: 5. iNNOVATIONfORUMfürMikrotechnik : 27. Februar 2013, Villingen-Schwenningen. Villingen-Schwenningen (2013), Poster, 1 S

Precision Handling of Electronic Components for PCB Rework

Gianmauro Fontana[1], Serena Ruggeri[1], Giovanni Legnani[2], and Irene Fassi[1]

[1] Institute of Industrial Technologies and Automation, CNR, Milan, Italy
{gianmauro.fontana,serena.ruggeri,irene.fassi}@itia.cnr.it
[2] Department of Mechanical and Industrial Engineering, University of Brescia, Brescia, Italy
{giovanni.legnani}@ing.unibs.it

Abstract. The paper focuses on the study of strategies and tools to handle miniaturized components in the electronic industry. In particular, the paper presents an innovative device and method to manipulate microcomponents by vacuum. The device includes an original releasing system, that does not require any external actuation, to assist their release. Indeed, at the microscale, adhesion forces predominate over the gravitational force due to the small masses of the microcomponents, often leading to the failure of the release phase if a release strategy is not implemented. The device, able to eliminate the adhesion problem, is compared with a traditional vacuum microgripper in terms of grasping and releasing error and percentage. The results of preliminary experimental tests are discussed, demonstrating that the innovative microgripper represents an interesting solution for handling electronic components as well as different microparts.

Keywords: Micro-handling, Micro-Robotics.

1 Introduction

In the last decades, a number of mechatronic devices have been developed for different purposes and have been broadening more and more in various fields. They result from a suitable integration of mechanical, information technology and electronic features that allow both to enhance the capabilities and the performance of standard products and to enable the conception and development of new generation systems facing the high-demanding requirements of an increasing market.

Products of daily usage such as cars, mobile phones and computers for example rely on electronics and contain Printed Circuit Boards (PCBs). They can mount few or hundreds components such as ball grid array (BGA) packages for integrated circuits, resistances or capacitors, often of tiny size. Indeed, a heavy trend to miniaturization has recently appeared, reflecting the need of getting smaller final products with integrated functionalities.

Despite the manufacturing technologies for electronic components and PCBs appear consolidated, few attention has been paid in the years to the development of suitable techniques to repair, reuse or recycle this kind of products when a malfunctioning or a breakdown occur. Most of times, due to economic reasons, the PCB is substituted with a new one and disposed, wasting many parts that could still have value and be recovered.

S. Ratchev (Ed.): IPAS 2014, IFIP AICT 435, pp. 52–60, 2014.
© IFIP International Federation for Information Processing 2014

However, it has been estimated that, from an environmental point of view, remanu-facturing is 80% more energy efficient than traditional manufacturing and, from an economic point of view, 60% more cost efficient [1].

If a repair plan is expected, the rework of PCBs is often performed manually, or in few cases with highly dedicated machines, and can result in a very expensive and complex operation; in the same way, the recovering of components in case of irreparable PCBs is not efficient.

According to the emerging "De-manufacturing" paradigm, innovative approaches and methodologies should be implemented to improve the management of electronic products at their end of life [2].

Consider for example the recovery of a working BGA from a defective PCB. In this case, the BGA is de-soldered from the PCB, by heating and consequently melting the soldering balls (with diameter <1mm). Then, a procedure called "re-balling" is needed to prepare the chip for the re-use, consisting in attaching new solder balls on the bottom surface of the BGA support. Nowadays, this operation is performed mainly manually with the help of fixtures and stencils or using preforms [3]. Specific stencils or preforms have to be used for each particular BGA support type, then limiting the flexibility of the process, and the operator ability influences the accuracy of the result. Recently, automatic rework stations have been developed [3], but still there is a need for more flexible and cost efficient automated equipment.

In this context, the present paper focuses on the study of methods and tools for handling miniaturized components to move towards a flexible and efficient rework of PCBs. The study is devoted to two main related aspects: the re-balling of BGA packages and the replacement of different surface-mount components (SMCs) on the PCBs. To overcome the limits of the current procedures, the work addressed the development of simple and low-cost gripping devices and strategies for both a complete or selective re-balling, then able to pick and place a whole grid of small solder balls or a single ball precisely and reliably. In the same way, the tools should be able to accomplish the replacement of SMCs of different size and mounted on the PCB with a generic orientation.

In particular, an innovative device and method to manipulate by vacuum a component has been conceived and prototyped. A patent application has been filed in March 2013 [4]. This device has been compared with a conventional vacuum microgripper in the execution of pick and place tests of both solder balls of various diameter and resistances.

The investigation of their performance considered both the gripping and the release phases. Indeed, at the microscale, due to the predominance of adhesion forces (e.g. capillary, electrostatic and Van der Waals forces) over the gravitational force, the release of microcomponents is often prevented or uncertain. In literature many solutions have been proposed to cope with this issue [5], but the search of effective and efficient methods is still an undergoing study. For this reason, a special releasing system has been designed and integrated in the innovative gripping device to release the components precisely, reliably and safely, avoiding an excessive increase in weight of the device itself or complexity of its structure.

The two types of handling devices are presented in Section 2. Section 3 describes the experimental tests and the performance indices considered for the comparison and discuss the preliminary results, highlighting the benefit of the proposed solution.

2   The Handling Devices

A convenient solution for the manipulation of electronic components is represented by the use of vacuum grippers, commonly used for macrocomponents and here adapted to the tiny size of the microcomponents. They are based on the pressure difference between the gripper and the atmosphere and have a generally simple structure, since they are mainly based on a small suction hose; they are not expensive and can be used to manipulate a wide range of components, even fragile.

In this work, two types of vacuum microgrippers were considered. The former is a standard microgripper: it is a commercially available needle for dispensing, basically consisting of a cannula connected to a vacuum generation system. The cannula has an internal diameter of 260 µm, which is attached to the hollow needle body that can be connected to the end-effector of a robot through a proper mechanical interface.

However, in [6], it has been shown that the release is greatly affected by the presence of the adhesion forces and that the simple switch off of the vacuum is often not sufficient. Therefore, suitable expedients need to be found. For example, the adhesion due to the electrostatic force can be reduced coating the glass pipette for the suction with a conductive layer connected to the ground. Moreover, the microcomponent could be guided to hit a sharp edge or rolled on the release plane. The inertial force of the component, generated by moving the gripper brusquely upwards for a short stroke, can also be exploit to ease the release, although results can be inaccurate [6]. Finally, the application of positive pressure, that is of a soft blow for some milliseconds, allows the achievement of a high release percentage and does not affect negatively the final precision of positioning. However, the tuning of the parameters related to the blow (e.g. its intensity) could sometimes be difficult and a trade-off between percentage and precision of release of the specific component has to be found.

On the basis of the results obtained in past experiments and the study of the detected behaviours, a new vacuum micro-gripping tool (No. 1 in Fig. 1) has been conceived and developed, able to cope with the micropart release issues. Similarly to a standard vacuum microgripper, this device is based on the pressure difference between the gripper and the atmosphere and basically consists of a cannula or suction hose connected to a vacuum generation system for picking a component by suction on the gripping end. It integrates an innovative mechanical system to assist the release phase of micro components precisely, reliably and safely, avoiding a considerable increase in weight of the device itself or an excessive complication of the system. In details, the mechanical system is inserted at least partially in the manipulation device and movable from a release position where a release portion projects externally from the gripping end, and a gripping position where the release portion returns into the manipulation device. The mechanical system includes a transversal extension, that in the current case is a holed disc, and a needle attached to the disc and having a diameter smaller than the internal diameter of the cannula. The needle is inserted at least partially inside the cannula and includes the release portion.

The mechanical release system is designed to be moved from the release position to the gripping position through activation of the vacuum generating system, which sucks in the mechanical release system in opposition to a return force pulling it toward the release position, for example the force of gravity acting on the mechanical release system. It is also designed to be moved from the gripping position to the release position by the said return force following a reduction or elimination of the pressure difference between the inside and outside of the manipulation device operated by the vacuum generating system.

A main feature consists in using the actuation principle for the part picking to move the release system, without the need of additional actuators, that would make the system more complex, heavier, bigger and more expensive.

3   Preliminary Experiments and Results

In this section, the results obtained through a set of preliminary experiments are presented. The tests evaluated the performance of the two types of grippers both in gripping and releasing different components. In particular, the handling of solder balls with diameter of 500 microns and mass of 0.00055 g, and SMD resistances with size 1.5 x 0.8 x 0.45 mm and mass of 0.002 g was evaluated. Each test consisted in the execution of a set of 30 standard pick and place cycles, representative of the real trajectory that the gripper with the component should perform for assembly or PCB rework.

The tests were carried out exploiting the setup available in our micro-manipulation work-cell [7]: a Mitsubishi Electric RP-1AH robot (Fig. 1) was used as motion system, while the measurements of the position and orientation of the microparts was obtained by a suitable vision system consisting of a camera combined with a macro lens, providing a bottom view of working (gripping and releasing) area with a spatial resolution of about 8.1 μm.

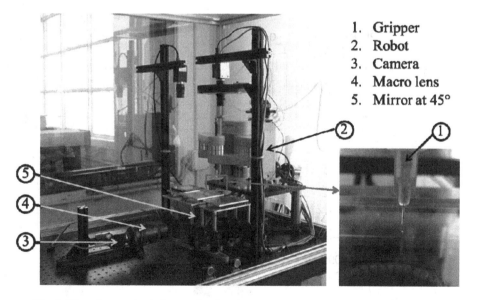

1. **Gripper**
2. **Robot**
3. **Camera**
4. **Macro lens**
5. **Mirror at 45°**

Fig. 1. The micromanipulation workcell and the innovative gripper with release needle

The gripping performance was evaluated comparing the measured position of the barycenter and the orientation of the component in two images acquired before and just after the gripping respectively. Similarly, the release was evaluated by detecting and measuring its pose before and after the release.

With the same approach adopted in [6], the performance indices we calculated for each test cycle were the repeatability and accuracy, according to the definition given by the international standard ISO 9283. Moreover, the gripping and release percentages, representing the success of the correspondent operations, were calculated.

As introduced above, in previous studies we demonstrated that the use of the positive pressure, in practice a soft blow, represented an acceptable method to assist the microcomponent release.

For this reason, the first class of experiments we executed considered the use a standard vacuum gripper, grasping the component by simple activation of the vacuum generation system, but releasing it by providing a blow for few milliseconds. A set of pick and place cycles was performed. While the gripping of the ball was always successful (gripping percentage of 100%), the release phase highlighted a strong dependence of the release performance on the blow pressure. Therefore two sets of experiments were carried out varying the blow pressure, the former trying to obtain an accurate and repeatable release independently from the release percentage, the latter trying to maximize the release percentage. The results for repeatability and accuracy for the first condition are graphically reported in Fig. 2, those related to the second condition are reported in Fig. 3. The origin of the reference system in the graphs represented the target position, while the small crosses indicated the barycenter of the ball for each cycle.

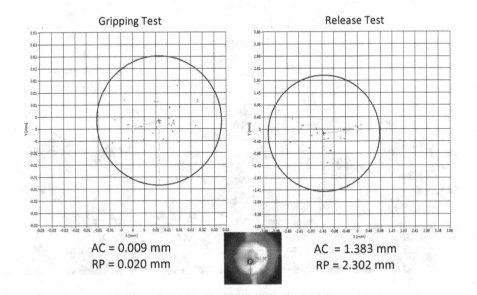

Fig. 2. Results for standard vacuum gripper manipulating the ball: gripping percentage of 100% and release percentage of 82.14%

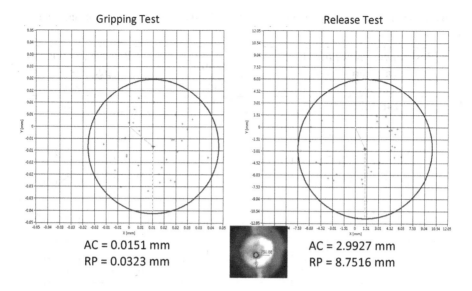

AC = 0.0151 mm
RP = 0.0323 mm

AC = 2.9927 mm
RP = 8.7516 mm

Fig. 3. Results for the standard vacuum gripper manipulating the ball: gripping percentage of 100% and release percentage of 86.67%

The experiments were repeated with the resistance and the results are shown in Fig. 4. Opposite to the ball, the resistance could have various orientations, therefore the accuracy and the repeatability values of the gripping and the release operations have been calculated to evaluate the orientation error and reported in Table 1.

As shown in Fig. 2-3, the percentage of release is not 100% while the accuracy and repeatability values are sometimes too high for ultra-precise manipulation. This is evident in Table 1, showing high values for both the orientation accuracy and repeatability.

The results in Fig. 4 also shows values of the gripping accuracy and repeatability for the resistance higher than those for the solder ball, most likely due to a sort of auto-centering property of the ball on the gripper tip. Moreover, the tuning of the pressure parameters can be onerous and the performance indices very sensitive to small changes of their values.

The innovative microgripper has been designed to combine the advantages of the vacuum gripping method and overcome all the difficulties related to the use of positive pressure, since no external actuation is needed to assist this phase. To investigate this gripper and the manipulation method, we repeated the experiments on the solder ball.

The gripping and release accuracy and repeatability are reported in Fig. 5. In this case, the performance indices for the release are much better than those obtained with the standard gripper and the positive pressure assisting the release. Note also that, in both cases, the percentage of success of the gripping and release operations was 100%, validating the expected effectiveness of this method.

Table 1. Orientation accuracy and repeatability for the standard microgripper manipulating the resistance

Resistance: values for orientation	Gripping	Release (by positive pressure)
Accuracy [°]	-0.550	-8.331
Repeatability [°]	17.293	43.322

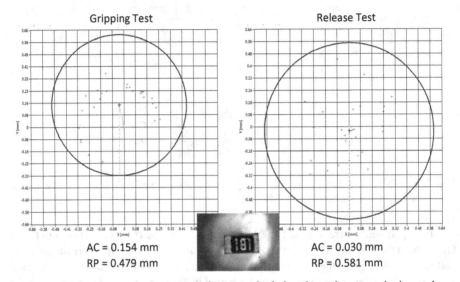

AC = 0.154 mm
RP = 0.479 mm

AC = 0.030 mm
RP = 0.581 mm

Fig. 4. Results for the standard vacuum gripper manipulating the resistance: gripping and release percentage of 100%

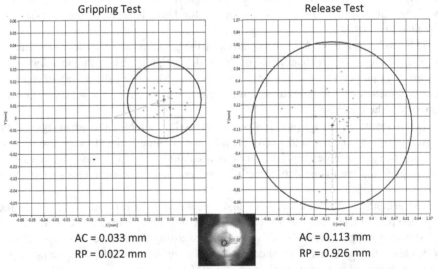

AC = 0.033 mm
RP = 0.022 mm

AC = 0.113 mm
RP = 0.926 mm

Fig. 5. Results for the innovative vacuum gripper manipulating the solder ball: gripping and release percentage of 100%

The experiments were then repeated with the resistance and the results are shown in Fig. 6. Note that the performance of the two microgrippers in terms of both precision and reliability are comparable. As for the case of the standard microgripper, the accuracy and the repeatability values of the gripping and the release operations have been calculated to evaluate also the orientation error (Table 2).

Table 2. Orientation accuracy and repeatability for the innovative microgripper manipulating the resistance

Resistance: values for orientation	Gripping	Release
Accuracy [°]	0.792	2.478
Repeatability [°]	10.498	45.759

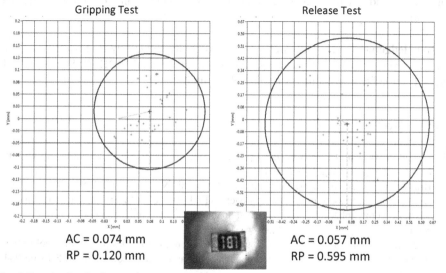

Gripping Test   Release Test

AC = 0.074 mm   AC = 0.057 mm
RP = 0.120 mm   RP = 0.595 mm

Fig. 6. Results for the innovative vacuum gripper manipulating the resistance: gripping and release percentage of 100%

4   Conclusions

Considering the current need for more flexible and cost efficient automatic PCBs rework stations to move towards the actual implementation of the de-manufacturing paradigm, the present work devoted to the study of handling devices for the re-balling of BGA packages and PCBs and the replacement of different surface-mount components (SMCs) on the PCBs.

An innovative device and method to manipulate by vacuum a microcomponent has been presented and compared with a conventional vacuum microgripper in the execution of grasping and releasing tests of electronic components. In particular, the release of the microcomponents, often prevented or uncertain with traditional handling devices due to the presence of adhesive forces, has been investigated. The results demonstrated that the innovative microgripper is able to manipulate submillimetric components better than the conventional vacuum microgripper and that the special releasing system allowed a reliable and safe release. However, the values of orientation accuracy and repeatability are still high for ultra-precise manipulation, therefore additional expedients should be considered. Moreover, it can be noticed that, for both microgrippers, the repeatability values in the gripping and release phases are much higher than the accuracy values. For this reason, further efforts will be required to better comprehend this phenomenon and try to improve the grippers' performance.

Concluding, the innovative device represents an interesting solution for handling electronic components for PCBs rework and, more in general, for micromanipulation and assembly of different microproducts.

In the next future, the handling of other microcomponents with this gripper will be also tested and it will investigated more deeply in order to highlight advantages and limitations.

Aknowledgements. This research has been partially funded by Regione Lombardia, in the framework of the Accordo Quadro RL-CNR, project 'FIDEAS: Fabbrica Intelligente per la Deproduzione Avanzata e Sostenibile'.

References

1. Errington, M., Childe, S.: A business process model of inspection in remanufacturing. Journal of Remanufacturing 3, 1–22 (2013)
2. Brusaferri, A., Colledani, M., Copani, G., Pedrocchi, N., Sacco, M., Tolio, T.: Integrated De-manufacturing systems as new approach to End-of-Life management of mechatronic devices. In: 10th Global Conference on Sustainable Manufacturing, pp. 332–339 (2012)
3. Solderquik, http://www.solderquik.com
4. Ruggeri, S., Fontana, G., Fassi, I., Legnani, G., Pagano, C.: Dispositivo di manipolazione e metodo per manipolare a vuoto un componente. Italian Patent pending No. MI2013A000541 (2013)
5. Chen, T., Sun, L., Chen, L., Rong, W., Li, X.: A hybrid-type electrostatically driven microgripper with an integrated vacuum tool. Sensors and Actuators A: Physical 158(2), 320–327 (2010)
6. Ruggeri, S., Fontana, G., Pagano, C., Fassi, I., Legnani, G.: Handling and Manipulation of Microcomponents: Work-Cell Design and Preliminary Experiments. In: Ratchev, S. (ed.) Precision Assembly Technologies and Systems. IFIP AICT, vol. 371, pp. 65–72. Springer, Heidelberg (2012)
7. Fontana, G., Ruggeri, S., Fassi, I., Legnani, G.: Flexible vision based control for microfactories. In: Proc. of the 7th International Conference on Micro- and Nanosystems IDETC/MNS 2013, Portland, OR, USA, August 4-7 (2013)

Shift Dynamics of Capillary Self-Alignment

Gari Arutinov[1,2,*], Massimo Mastrangeli[3], Edsger C.P. Smits[1], Gert van Heck[1], Herman F.M. Schoo[1], Jaap J.M. den Toonder[2], and Andreas Dietzel[4]

[1] Holst Center/TNO, High Tech Campus 31, 5656 AE Eindhoven, The Netherlands
{gari.arutinov,edsger.smits,gert.vanheck,herman.schoo}@tno.nl
[2] Eindhoven University of Technology, Microsystems, 5600 MB Eindhoven, The Netherlands
{j.m.j.d.toonder@tue.nl}
[3] BEAMS, Université Libre de Bruxelles, Av. Fr. Roosevelt 50, B1050 Bruxelles, Belgium
{massimo.mastrangeli@ulb.ac.be}
[4] Technische Universität Braunschweig, Institut für Mikrotechnik,
38124 Braunschweig, Germany
{a.dietzel@tu-braunschweig.de}

Abstract. This paper describes the dynamics of capillary self-alignment of components with initial shift offsets from matching receptor sites. The analysis of the full uniaxial self-alignment dynamics of foil-based mesoscopic dies from pre-alignment to final settling evidenced three distinct, sequential regimes impacting the process performance. The dependence of accuracy, alignment time and repeatability of capillary self-alignment on control parameters such as size, weight, surface energy and initial offset of assembling dies was investigated. Finally, we studied the influence of the dynamic coupling between the degenerate oscillation modes of the system on the alignment performance by means of pre-defined biaxial offsets.

Keywords: capillarity, self-alignment, dynamics, fluidics, packaging.

Surface tension-driven self-alignment (SA) is a simple, accurate and cost-effective technique for heterogeneous integration and stacking of dies onto pre-patterned substrates. A vast class of fluidic self-assembly processes was thoroughly investigated and developed since the early 90's, and many applications were successfully demonstrated [1-2]. Capillary SA rapidly emerged as a remarkable way to overcome the accuracy/throughput trade-off limiting established techniques of pick-and-place microassembly [3]. It indeed combines the manipulative dexterity of robotic handling for fast and rough component pre-alignment with the highly-accurate and passive final alignment yielded by the relaxation of a liquid droplet over a shape-matching patterned confinement site [4]. Die fetching and registration can thus be partly parallelized, leading to increased efficiency in the integration of innovative microsystems [5]. A more comprehensive knowledge of capillary SA needs further modeling and experimental testing. Within the past two decades researchers made significant progress in the former direction [6-10]. However, most of modeling works are based on quasi-static simulations, whereas the dynamics of the process has been rarely tackled so far.

---

[*] Corresponding author.

S. Ratchev (Ed.): IPAS 2014, IFIP AICT 435, pp. 61–68, 2014.

In this paper we present an experimental study of the in-plane dynamics of capillary SA. We used transparent mesoscopic dies as components to be assembled onto shape-matching receptor sites patterned onto the substrates. Using water droplets in air and following the experimental procedure described in Fig. 1, mm- to cm-sized square foil dies could be aligned onto the patterned carriers with accuracy better than 30 µm, *i.e.* with high accuracy relatively to their lateral dimensions, and comparable to dicing tolerances. We analyzed the full uniaxial dynamics of capillary SA from die pre-alignment to final settling, evidencing the presence of three distinct, sequential regimes and highlighting their impact on the process performance. The dependence of accuracy, alignment time and repeatability of capillary SA on control parameters such as size, weight and surface energy of assembling dies, and their initial misalignment was additionally investigated. Finally, we studied the influence of the coupling between the degenerate in-plane oscillation modes of the dynamic system on the alignment performance by means of pre-defined biaxial offsets.

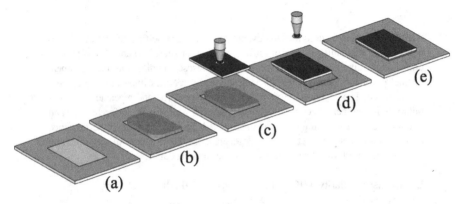

Fig. 1. Sketch of the process steps for capillary self-alignment of a foil die: (a) patterning of the carrier substrate, (b) deposition of a droplet of assembly liquid, (c) coarse alignment of a functional die, (d) the liquid wets foil die forming a meniscus, (e) self-alignment of a foil die on the corresponding binding site.

Foil dies were fabricated from transparent 125-µm-thick Polyethylene Naphthalate (PEN) sheets (Teonex® Q65FA, DuPont®). Using a frequency-tripled Nd:YAG laser, a PEN sheet was diced into a set of square-shaped foil dies ranging in dimensions from 2 to 20 mm with dicing accuracies down to 30 µm. To quantify alignment dynamics and placement accuracy, marker structures were engraved in the center of each foil die (Fig. 2b). The laser-ablated PEN foils were aligned onto Au-patterned Si monitor wafers (Fig. 2a), allowing for reproducible and stable surface conditions in the experiments (Fig. 3). The dies were handled through a homebuilt micropositioning stage equipped with integrated vacuum tweezers (SMD-VAC-HP, Vacuum Industries Inc.). A positioning base stage (XYZ 500 TIS, Quater Research and Development) providing 10 µm-resolved displacements along three axes was used to pre-align the dies. High-speed camera stage and image recognition software were used to track and analyze the process dynamics.

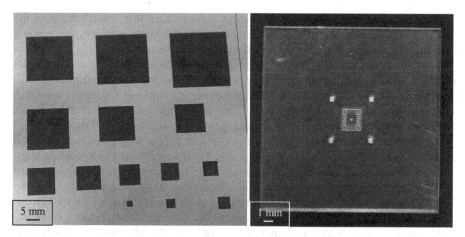

Fig. 2. (a) Optical image of a Si wafer with top patterned Au layer functionalized with perfluoro-decanethiol SAM. Dark square areas are hydrophilic SiO_2 binding sites. Marker structures consisting of 50 μm-wide Au lines are patterned at the center of each site. (b) Optical image of a transparent, 125-μm-thick PEN die cut by a frequency-tripled Nd:YAG (355 nm, 25 ns) laser. The laser was also used to engrave marker structures at the center of the foil die. The combined markers allow registering the position of the foils with accuracies down to 20 μm.

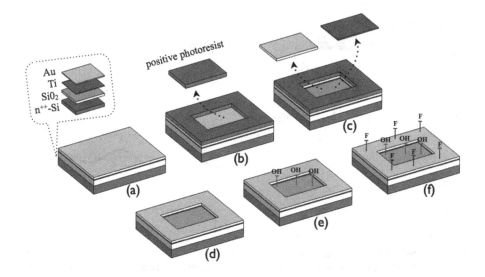

Fig. 3. Schematics of the patterning process for the Si substrate. (a) Ti (5 nm) and Au (100 nm) layers are sequentially sputtered over a 1 μm-thick layer of SiO_2 thermally-grown on the Si wafer. (b) Positive photoresist is spin-coated, patterned and developed to define the binding sites. (c) Au and Ti are removed from the binding sites by wet etching. (d) The remaining photoresist is striped. (e) The substrate is cleaned by O_2 plasma. (f) Selective deposition of a fluorinated alkanethiol self-assembled monolayer on the Au surface. Relative dimensions out of scale for illustration purposes.

Analysis of recorded SA trajectories (Fig. 4a) revealed that, upon release of the foil die from a relatively-large uniaxial offset from the target position, the ensuing dynamics of capillary SA can be divided into three separate regimes [11]. *Transient wetting* comes first, whereby the foil stays initially at rest after landing on the droplet (Fig. 4a). After contact with the bottom surface of the foil die, the liquid forms a bridge and spreads across the slit geometry. The relaxation of the liquid meniscus continues till it reaches and pins over at least one of the boundaries of the foil die. This event marks the beginning of the uniaxial translational motion of the foil die. In this second regime the foil die moves with a constant acceleration (see inset in Fig 4a) and hence describes a parabolic trajectory. It is therefore subjected to a constant lateral force, which arises from the deformation of the meniscus and the gain in surface energy as surface wetting progresses. This *constant acceleration* regime fades as the die reaches the vicinity of the target position and dissipative viscous forces become comparable with the driving capillary forces. The final dynamic regime can be described by a *damped harmonic oscillation* (DHO, Fig 4a) observed in flip-chip assembly, as well [6-7]. By tailoring surface wettability, we contextually showed that transient wetting and constant acceleration regimes are strongly dependent on and can be tuned by the surface energy of the bottom surface of the assembling foil dies (Fig. 4b). Particularly, increased surface energy—*i.e.*, higher hydrophilicity and lower contact angle—decreases the total time-to-alignment.

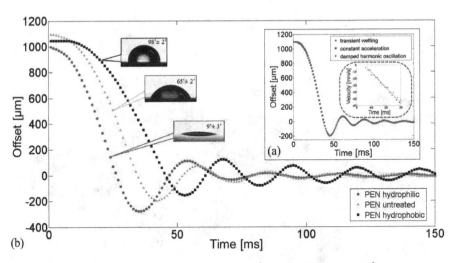

Fig. 4. (a) Self-alignment trajectory of 18×18 mm^2 foil die (0.80 mg/mm^2) extracted from self-alignment process recorded by high-speed camera. (inset) Numerical derivative of the parabolic regime depicted in (a), evidencing linear progression of velocity in time. (b) Self-alignment trajectories of 18×18 mm^2 PEN foil die (0.80 mg/mm^2) with no coating and hydrophilic and hydrophobic coating.

In contrast to the sub-millimetric case, the dies used hereby had a Bond number $Bo = \rho g L^2/\gamma \gg 1$. Hence the inertia of the considered mesoscopic foil dies was expected to significantly affect their self-alignment dynamics. Sizes and weights of foil dies were thus investigated to find a reliable process window. Foil dies were coarsely aligned manually with an initial offset of 20% of their lateral size, preventing offset-dependent alignment failure [12]. The water layer thickness (125 µm) was the same for all binding sites. As seen in Fig. 5, the assembly yield was substantially lower for the smallest and the largest foils. Moreover, for the smallest foils the accuracy of the initial offset was hardly reproducible (standard deviation of about 30 µm). Consequently, the SA process was considered unreliable.

To investigate the influence of inertia on the DHO regime, equal-sized PEN dies with mass densities varying from 0.18 to 0.98 mg/mm^2 were pre-aligned over corresponding binding sites with prefixed initial offset of 950 ± 50 µm. Their SA trajectories evidenced a net impact of inertial effects on the oscillation period (Fig. 6).

We additionally investigated the SA dynamics for quasi-1D cases. A foil die was repeatedly released at different, pre-fixed small offsets along the secondary axis while the initial offset along the principal axis was kept constant within loading precision. The corresponding SA data revealed no divergence of the trajectories along the principal axis during the constant acceleration regime (Fig. 7a) in spite of significant differences in the dynamics along the secondary axis (Fig. 7b). Hence we derive that non-negligible yet relatively-small positioning errors along the secondary axis barely affect the lateral alignment dynamics of square foil dies along the principal axis.

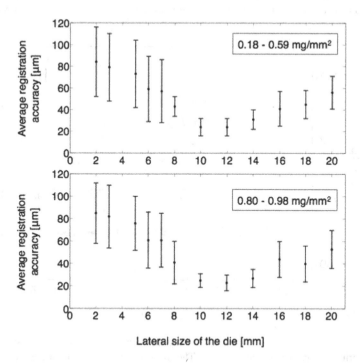

Fig. 5. Average registration accuracy for self-aligned foil dies of varying sizes and weights. Foil dies of each size and weight were self-aligned at least ten times from an initial offset of about 20% of foil die lateral size. A 125µm-thick layer of water was used in all cases.

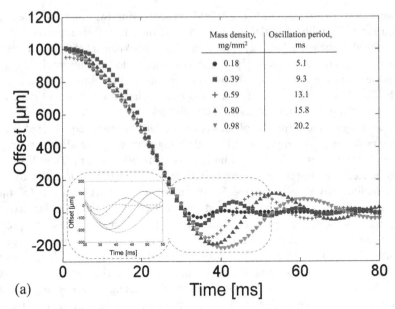

Fig. 6. Self-alignment trajectories of foil dies with different mass densities (see inset), and corresponding oscillation periods as extracted from experimental trajectories (shown in inset)

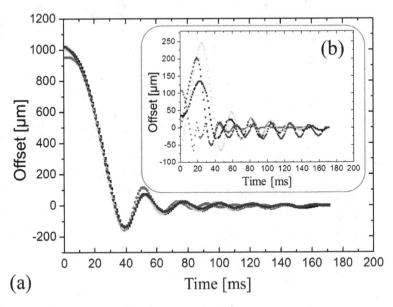

Fig. 7. Self-alignment trajectory of a foil (0.80 mg/mm²) die as recorded by high-speed camera along (a) principal and (b) secondary axis. The die was repeatedly released at various initial offsets along the secondary axis, while initial offsets along main axis was kept the same within loading precision of the micropositioner.

As a generalization of the previous case involving weak coupling between the degenerate in-plane oscillation modes, we analyzed the SA dynamics for the case of initial die offset along the main diagonal direction (*i.e.* 45 deg)—*i.e.* for equal, relatively-large initial die offsets along both axes, and consequent strong coupling between the modes. Recorded SA trajectories evidence an initially linear translation of the die along the diagonal direction, followed by an inward spiraling trajectory as the die ultimately approaches the center of the target site (Fig. 8a). A DHO equation provided also in this case a very good fitting for the projection of the oscillatory SA trajectory along the main axes (Fig. 8b). Final accuracy for diagonal self-alignment matched the best uniaxial case. Notably, for a given, small radial offset from the center of the binding site, the capillary restoring force ensuing from a diagonal displacement is larger than for a uniaxial displacement. This may make diagonal alignment preferable to the uniaxial, as it may avoid issues of the latter for small offsets [12].

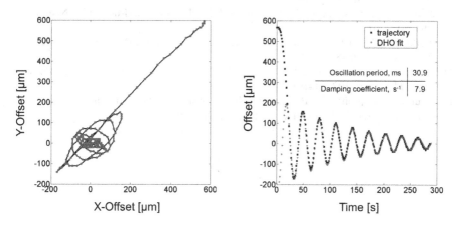

Fig. 8. (a) The phase plane of self-alignment trajectory of a foil (0.80 mg/mm^2) die as recorded by high-speed camera along diagonal axis, and (b) self-alignment trajectory fitted with damped harmonic oscillation (DHO) approximation and corresponding oscillation period and damped coefficient as extracted from trajectories (shown in inset). The die was released along the diagonal axis, where initial offsets along secondary and principle axes were in principle equal.

In conclusion, our investigation of the dynamics of capillary SA significantly extends the understanding of the process available in literature, and suggests additional parameters to control its performance, such as the optimal uniaxial offset window and the surface energy of the die. Though our analysis focused on millimetric dies, most of the results can directly be applied to the case of much smaller dies—in fact, the use of larger beside transparent vehicles just makes the phenomena more apparent with respect to the microscopic scale, thus easier to detect and study. Finally, this new body of knowledge can further ease or extend the compliant adoption of the technique for industrial microsystem manufacturing.

References

1. Mastrangeli, M., Abbasi, S., Varel, C., van Hoof, C., Celis, J.-P., Böhringe, K.F.: J. Micromech. Microeng. 19, 081001, 37 p. (2009)
2. Crane, N.B., Onen, O., Carballo, J., Ni, Q., Guldiken, R.: Microfluid. Nanofluid. 14, 383–419 (2013)
3. Morris, C., Stauth, S.A., Parviz, B.A.: IEEE Trans. Adv. Pack. 28, 600–611 (2005)
4. Sariola, V., Jääskeläinen, M., Zhou, Q.: IEEE Trans. Robot. 26, 965–977 (2010)
5. Mastrangeli, M.: Surface tension-driven self-assembly. In: Lambert, P. (ed.) Surface Tension in Microsystems, ch. 12, pp. 227–253. Springer (2013)
6. Lin, W., Patra, S.K., Lee, Y.C.: IEEE Trans. Compon. Packag. Manuf. Technol. 18, 543–551 (1995)
7. van Veen, N.: J. Electron Packag. 121, 116–121 (1999)
8. Mastrangeli, M., Valsamis, J.-B., van Hoof, C., Celis, J.-P., Lambert, P.: J. Micromech. Microeng. 20, 075041 (2010)
9. Lambert, P., Mastrangeli, M., Valsamis, J.-B., Degrez, G.: Microfluid. Nanofluid. 9, 797–807 (2010)
10. Berthier, J., Mermoz, S., Brakke, K., Sanchez, L., Frétigny, C., Di Cioccio, L.: Microfluid. Nanofluid 14, 845–858 (2013)
11. Arutinov, G., Mastrangeli, M., Smits, E.C.P., Schoo, H.F.M., Brugger, J., Dietzel, A.: Appl. Phys. Lett. 102, 144101 (2013)
12. Arutinov, G., Smits, E.C.P., Mastrangeli, M., van Heck, G., van den Brand, J., Schoo, H.F.M., Dietzel, A.: J. Micromech. Microeng. 22, 115022 (2012)

Image Stitching Based Measurements of Medical Screws

Zoran Cenev, Timo Prusi, and Reijo Tuokko

Department of Mechanical Engineering and Industrial Systems,
Tampere University of Technology, Finland
{timo.prusi,reijo.tuokko}@tut.fi, zoran.zenev@aalto.fi

Abstract. Image stitching is a method that forms a bigger image out of two or more smaller images with a certain overlap in their field-of-view. Usage of this technique is present in photography, medical imaging and other fields as well. The potential for using image stitching for measurement purposes in industrial applications has not been investigated as thoroughly, but it is receiving more attention. This paper examines the possibility for creating a machine vision-based measurement system that employs commercially available image-stitching platforms for measuring the length of long medical screws. Three methodologies were tested for performing medical screw measurements and this paper summarizes the findings of the image-stitching approach.

Keywords: image stitching, measurements, machine vision.

1 Introduction

It is well known that screws are manufactured in lot sizes. The common inspection of these mechanical fasteners is by trial and error i.e. pushing them through thread gauges. Often, if a screw fails the inspection process, they are disposed of and recycled for another use. Unlike mechanical screws, medical screws (see **Fig. 1** below) are built from expensive materials [1] which cannot be recycled if they are faultily produced. Therefore, the inspection process has to be accurate enough in order to support the fine-tuning of the screw's manufacture. One very important aspect of medical screws that necessitates their high-quality inspection is the fact that they enter, stay and degrade within the human body. As a result, the quality regulations in the United States of America (stipulated by the Food and Drug Association, the FDA) and in Europe (overseen by the European Medicines Agency, EMEA) are extremely rigorous.

This paper introduces three machine vision approaches to the inspection of a medical screw as shown in **Fig. 1**. The inspections process is based on measuring certain screw parameters, such as the thread inner and outer diameter, the screw's total length, and the diameter of the neck (the part between the screw head and the beginning of the thread), and verifying that they are within given tolerances.

Currently, these measurements are done using a manual caliper. However, this way of measuring is very inefficient when measuring a batch of screws. For example, this approach is prone to human errors and features low reliability, paper-based data and inadequate opportunities for automation. Therefore building a machine vision-based

S. Ratchev (Ed.): IPAS 2014, IFIP AICT 435, pp. 69–78, 2014.

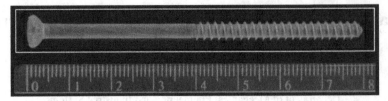

Fig. 1. Medical screw. White rectangle indicates field-of-view of 90 mm x 120 mm.

measurement system is regarded as a very valuable approach to the problem and has been researched in great detail.

Dimensions such as the thread's inner and outer diameter, and the diameter of the neck can be inspected with any machine vision set-up that contains an ordinary industrial camera equipped with a high quality lens that captures a field-of-view of a few dozen millimeters. However, the real challenge arises when measuring the screw length had to be realized. Hence, the focus of this paper is on reporting the results and analysis of an image-stitching based metrology approach for measuring the screw length.

Applications of the image-stitching technique in metrology were reported in the articles [2][3][4][5]. According to the findings reported in [2] and [3], image stitching was successfully implemented in performing AFM (Atomic Force Microscope) linewidth measurements through the manual stitching of two images. The article in [4] reports achieving significant similarities in accuracy of the linear measurements of stitched compared with non-stitched imaging in cone beam computed tomography. Commercial use of image stitching metrology in generating high-resolution nanoscaled topography maps is reported in [5].

2   Screw Measurement Approaches

A medical screw 80 mm in length is manufactured with a length tolerance of 0.5mm. All the cross-sectional measurements, such as the thread outer/inner diameter and the neck diameter are manufactured with a tolerance of 0.05mm. Nielsen [6, page 3] suggests that the measurement resolution should be at least one fourth of the smallest manufacturing tolerance, i.e. 0.0125 mm in this case.

A field-of-view of about 90 mm x 12 mm was desired in order to look at all the component dimensions and allow some margins for part positioning (see **Fig. 1**).

The first approach was a solution in which all measurements could be done from one single image. Therefore, the overall measurement resolution in millimeters was the smallest required resolution from all the previously mentioned ones, i.e. 0.0125 mm. This is the smallest feature that needed to be detected from the single image (at least 1 pixel in size) leading to a theoretical spatial resolution of 1 pix / 0.0125 mm or, other way around, 80 pix / mm. However, according to the Nyquist-Shannon theorem [7], [8], the spatial resolution has to be at least two times greater than the calculated ratio, i.e. 160 pix / mm. Moreover, a rule of thumb in machine vision systems is that

the spatial resolution should be 4 to10 times greater than the calculated ratio. Therefore a spatial resolution of 320 pix / mm was desired.

In order to cover the above mentioned FOV with the calculated spatial resolution,
we needed an image of size:

$$width_{pixels} = 90 \; mm * 320 \; \tfrac{pix}{mm} = 28\,800 \; pix \tag{1}$$

$$height_{pixels} = 12 \; mm * 320 \; \tfrac{pix}{mm} = 3\,840 \; pix \tag{2}$$

In reality, camera sensors with this kind of aspect ratio (width/height) are not
available. Common aspect ratios are in the range of 4:3. Assuming an image height of
3840 pix and an aspect ratio of 4:3, the image width would be only 5120 pix covering
only 16 mm of the length of the screw. Or, to put it another way, if we want a 28 800
pixel wide image, the image height would be (with 4:3 aspect ratio) 21 600 pix, requiring about 622 megapixel camera covering a FOV of 90 mm x 66.5 mm. This approach demands an extremely high resolution camera which make it very expensive,
and therefore impractical for this case.

Another approach was to apply the Coordinate Measuring Machine CMM method.
This means that the screw can be positioned on a linear axis and the camera can capture images from both ends of the screw. The length can be derived from the distance
measured from the movement of the linear axis combined with the differences between the screw ends of the images. This approach is not the focus of this paper, and
will not be examined in further detail.

Third approach was image-stitching-based metrology, and this approach is described in the text below.

Image stitching is a process that takes multiple images with overlapping field-of-
views and outputs a panoramic image. As illustrated in **Fig. 2**, the panoramic image
on the far right is obtained by stitching the two input images on the left. It is evident
from **Fig. 2** that the input images have a certain FOV overlap [9].

Image Input 1   **Image Input 2**   **Stitched Image**

Fig. 2. The Image-Stitching Process

The idea behind this approach is to capture certain amount of images along the
medical screw, stitch them together and then perform measurements from the obtained panoramic image.

3   Machine Vision Setup

3.1   Physical Components

On the left of **Fig. 3** one can see the machine vision setup. This consists of a monochromatic industrial camera (2560 x 1920 pixel resolution) equipped with a telecentric lens giving a field-of-view of 18 mm x 12 mm. This gives a spatial resolution of about 160 pix / mm, which theoretically satisfies the requirement for measuring the cross-sectional dimensions with the specified measurement resolution calculated in Chapter 2. However, this spatial resolution is only half of the desired one.

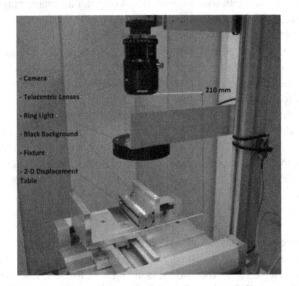

Fig. 3. Machine Vision System for Image Capturing (left) panoramic stitched image (right)

The fixture on which the screw resided was attached to a two-dimensional displacement table and was moving in the direction along the screw axis. Red ring light was used as a front illumination source, and it was positioned between the camera and the target.

A black background was placed between the screw and the fixture, **Fig. 3** (right). The role of the background was to enable high contrast and to introduce more features in order to ease the image stitching process. The remaining components (not visible on the image above) are the computer to which the camera was connected and a controller that was used to control 2D moving table.

3.2   Stitching Platforms

This work sees the image stitching process as a black box. This means that a certain input goes into the box and a certain output comes out of the box, whereas the examiner did not know what was happening inside. This meant that ready-made image stitching software platforms such as the "Photomerge" tool in Photoshop and Micro-soft's Image Composite Editor ICE were used.

There are two reasons why these platforms were chosen over any others. First, the role of the user was just to input the images that he or she wanted to stitch, and the stitching process was performed without any further user involvement. Second, these platforms were capable of working with many image formats.

A planar reposition function supported by both platforms was used in order to avoid any rotational and spherical effects onto the stitching process.

3.3   Image Stitching Concept Validation Experiment

The concept validation was done by stitching images in order to obtain a panoramic image (**Fig. 3**, right) and then test the accuracy and the repeatability of the length measurements. The testing procedure started by realizing the image acquisition. **Fig. 4** shows how the images were acquired and stored.

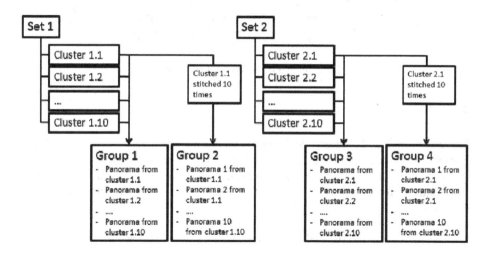

Fig. 4. Data buffering layout

The data from the image acquisition was divided into two sets, each having 10 clusters of images. Each cluster contained 11 images in the first set and 22 images in the second set; see **Fig. 4**. The amount of overlap required for the stitching process in the first and second set was ~60 % and ~80 % respectively.

The experiment continued by making panoramic images from each cluster using the two commercially available stitching platforms, Microsoft ICE and Adobe Photo-shop. Additionally, in order to verify the repeatability of the stitching process, clusters 1.1 and 2.1 were both stitched 10 times. As a result, each of the four groups of stitched images contains 10 panoramic images.

After the stitching phase of the experiment, the length of the screw in 40 stitched images was measured with a Clamp (Rake) tool in the National Instrument Vision

Assistant software platform. This tool finds the edges from rectangular region of interest (visible in **Fig. 5**) and calculates the distance between first and last edges found. Because of tool's operating principle, it is important to align the tool to the target being measured. When measuring screw lengths, alignment was assured by measuring screw orientation (deviation from horizontal) and then rotating the image so that the screw was horizontal before measuring the length with Clamp (Rake) tool. From the individual measurements, the average dimension, standard deviation and error range (max value – min value) were calculated. The repeatability of the Clamp (Rake) tool was estimated by recording 100 images of a known length metal piece (length 16 mm) that was small enough to fit in the field-of-view of a single image. **Fig. 5** shows the metal piece and the Clamp (Rake) tool. The results and analysis of these calculations are provided below.

Fig. 5. Screen capture of Vision Assistant showing the metal piece

4   Results

The error range and standard deviation of measuring the length of the metal piece from the 100 individual images were 0.03 % and 0.01 % respectively.

As **Table 1** shows, the error range and standard deviation in measuring the screw length ranged from 0.09 to 1.78 % and from 0.03 to 0.47 % of the screw length, respectively.

Table 1. Error range and standard deviation of the screw length obtained from the panoramic images

Stitching Platform	Clusters of 11 images (Group 1)	Clusters of 22 images (Group 3)
Microsoft ICE		
- Average Length [pix]	11 903	11 881
- Error Range [%]	0.10	0.09
- Standard Deviation [%]	0.04	0.03
Photoshop's photomerge		
- Average Length [pix]	11 930	11 935
- Error Range [%]	0.24	1.78
- Standard Deviation [%]	0.07	0.47

The best case was when the stitching process was performed with Microsoft ICE platform on a cluster from 22 images with a maximum error range of 0.09 % and a standard deviation of 0.03 %. Converting pixels to millimeters was done with calibrating the machine vision setup with National Instrument Vision Assistant calibration tools and laser marked calibration grid. In addition, the actual length of the screw was measured with a CMM. The obtained maximum error range was 0.072 mm and maximum standard deviation 0.024 mm.

4.1 Image Stitching Process Behavior

Since the image-stitching process was seen as a black box and since there was no possibility to have control over the stitching process, its behavior was examined. Four phenomena were identified: (1) Different image size for each panorama, (2) obtaining convex panoramic images, (3) unexpected stitching and (4) unrepeatable stitching.

Different Image Size
Table 2 shows the pixel size range (i.e. maximum width – minimum width and maximum height – minimum height) of each group of stitched images from both sets, Microsoft ICE and Photoshop.

Table 2. Image size range of the stitched images obtained from both sets

Image sets and stitching platforms	Image size range [pix]	
	Width (~ 13 000 pix)	Height (~ 2 000 pix)
Set 1 - Microsoft ICE		
Group 1	10	12
Group 2	1	5
Set 1 - Photoshop's photomerge		
Group 1	31	13
Group 2	55	6
Set 2 - Microsoft ICE		
Group 3	14	14
Group 4	2	0
Set 2 - Photoshop's photomerge		
Group 3	27	39
Group 4	30	18

Convex Panoramic Images
Another phenomenon that appeared was the bending effect, or convex panoramic image. This image is shown in **Fig. 6** and a similar image was obtained with both stitching platforms. The reason for this is unclear, because how the stitching platforms are "black boxes", i.e. the authors do not know how the stitching algorithm works. However, this effect was reduced by adding more features to the background, see **Fig. 3** (right). Therefore we can propose a hypothesis that the reason for this distortion is lack

of distinctive features along the screw, and especially in the upper and lower sections of the image (quite constant grey in **Fig. 6**). However, this was not tested.

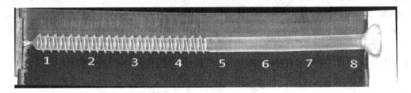

Fig. 6. Convex panoramic image

Unexpected Stitching
This phenomenon refers to undefined behavior of both stitching platforms and their abnormal outcome see **Fig. 7**.

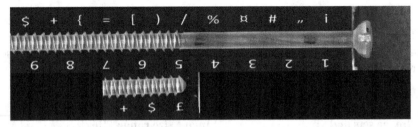

Fig. 7. Abnormal panoramic image

However, these kinds of images can be identified and thus deleted with a previously set image size threshold value.

Unrepeatable Stitching
Another undesired effect that was discovered from the stitching image process is the instability of the stitching platform itself. For example, the sizes of images obtained by stitching the same set of images differ by up to 55 pixels or 0.42 %, as shown in **Table 2** (see results for Groups 2 and 4). Naturally this also affects also the length measurements. **Table 3** shows the repeatability of the screw length measurement results of image groups 2 and 4 stitched with both platforms.

Table 3. Screw length deviations in panoramic images stitched from the same cluster of images

Stitching platform	Cluster of 11 images (Group 2)	Cluster of 22 images (Group 4)
Microsoft ICE		
- Error Range [%]	0.01	0.01
- Standard Deviation [%]	0.00	0.01
Photoshop's photomerge		
- Error Range [%]	0.56	1.09
- Standard Deviation [%]	0.18	0.27

5   Analysis and Discussion

The following findings can be identified from the results presented above:

- **Table 3** shows that Microsoft ICE is more repeatable than Photoshop. This is supported by **Table 2**, especially when looking at the results of groups 2 and 4.

- **Table 1** shows that also the length measurements have a smaller error range and standard deviation with Microsoft ICE than with Photoshop.

- Major concern is the fact that the average length varies (see **Table 1**) even though the images were captured from the same target with the same hardware setup, and the input images were the same for both stitching platforms.

- The software platforms' error range and standard deviation in **Table 2** are only affected by the repeatability of the image stitching process. In contrast, the results in **Table 1** are not only affected by image stitching process, but also by hardware and environment-related noise and unrepeatability. Therefore the best case in **Table 1** (Microsoft ICE and Set 2) has about three times greater error range and standard deviation than the measurements made from the metal piece which fitted into a single image FOV.

Since the smallest manufacturing tolerance was 0.05 mm, the resolution of the measurement system was expected to be a fourth of that, i.e. 0.0125 mm. Hence the obtained maximum error of 0.072 mm and the maximum standard deviation of 0.024 mm do not satisfy the desired criteria for length measurement with the desired cross-sectional measurement resolution. However, since the length manufacturing tolerance is exactly one order of magnitude higher than the one for cross-sectional dimensions, the measurement resolution can be increased for the same factor, resulting in a new length measurement resolution of 0.125 mm. Taking this into consideration, measuring the length can be achieved with the presented approach.

6   Conclusion

This paper asserts that measuring the length of long components such as a medical screw with a low-cost machine vision system with commercially available image stitching platforms is possible up to a certain level of measurement resolution.

Firstly, it should be remembered that the number of images used to create a panoramic image depends from the stitching environment. Microsoft ICE stitches marginally better with more input images; on the other hand Photoshop stitches much better with less.

Secondly, it is very likely that the stitching process can exhibit unbound and inconsistent behavior, such as the ones depicted in section 4.1.

Thirdly, the uncertainty from the stitching process is not the only source of uncertainties and errors in the actual measurements. The image acquisition and measurement algorithms also cause uncertainties in the measurements made from stitched images.

Fourth aspect is the interfacing and automation aspect. When building a measurement application utilizing a ready-made stitching platform, it should be as easy as possible to just input an image sequence and get a stitched image out. Therefore, the stitching platform should have an easy-to-use application protocol interface (API). Photoshop has this, but Microsoft ICE does not. Of course, building your own stitching platform is an option, but meeting the performance of Microsoft ICE would be challenging.

Acknowledgements. This work was done in a project funded by Finnish Funding Agency for Technology and Innovation (TEKES). Authors would like to thank TEKES and participating companies. In addition, authors are very grateful for the support from their colleagues Riku Heikkilä and Niko Siltala from the Microfactory research group at the Department of Mechanical Engineering and Industrial Systems.

References

[1] Törmälä, P., Vainioapää, S., Rokkanen, P.: U.S. Patent No. 5,084,051. Washington DC (January 28, 1992)

[2] Fu, J., Dixson, R., Orji, G., Vorburger, T., Nguyen, C.V.: Linewidth measurement from a stitched AFM image. International Conference on Characterization and Metrology for ULSI Technology. American Institute of Physics, Dallas (2005)

[3] Chu, W., Fu, J., Vorburger, T.V.: Subpixel image stitching for linewidth measurement based on digital image correlation. Measurement Science and Technology, 105104 (5pp) (2010)

[4] Srimawong, P., Krisanachinda, A., Chindasombatjaroen, J.: Accuracy of Linear Measurements in Stitched Versus Non-Stitched Cone Beam Computed Tomography Images. In: Challenges of Quality Assurance in Radiation Medicine, p. 40. Phitsanulok, Thailand (2012)

[5] Nanosurf Application Note (n.d.). Analyzing Large Surfaces using AFM Stitching. Liestal, Switzerland

[6] Nielsen, H.S.: ISO 14253-1 Decision Rules–Good or Bad. NCSL-I Workshop and Symposium, p. 7. Charlotte, NC, USA (1999)

[7] Olshausen, B.A.: Aliasing. PSC 129–Sensory Processes, 1-6 (2000)

[8] Candès, E.J., Wakin, M.B.: An Introduction To Compressive Sampling. IEEE Signal Processing Magazine, 21–30 (2008)

[9] Levin, A., Zomet, A., Peleg, S., Weiss, Y.: Seamless Image Stitching in the Gradient Domain. In: Pajdla, T., Matas, J(G.) (eds.) ECCV 2004. LNCS, vol. 3024, pp. 377–389. Springer, Heidelberg (2004)

Concept of a Virtual Metrology Frame Based on Absolute Interferometry for Multi Robotic Assembly

Robert Schmitt, Martin Peterek, and Stefan Quinders

Laboratory for Machine Tools and Production Engineering (WZL),
Chair of Production Metrology and Quality Management, RWTH Aachen University
{R.Schmitt,M.Peterek,S.Quinders}@wzl.rwth-aachen.de

Abstract. Highly individualized and customized products with dynamic life-cycles increase the need for flexible and reconfigurable assembly systems. Industrial robots are a key technology for future production systems especially for large scale components. The trade off between increasing work piece dimensions and constant or even increasing tolerance requirements, that are in some cases comparable to micro assembly systems, has to be solved by flexible and precise manufacturing and fixtureless assembly processes.

Keywords: Multi Line, virtual metrology frame, assembly.

1 Assembly of Large Scale Parts

The achievable accuracy of a robot based assembly system is influenced by the weight of work pieces and end-effectors, process forces, gravitation and temperature of the surroundings. Manufacturing, handling and assembly processes of large scale parts often take place in production sites with harsh environmental conditions. Temperature changes are influencing the handling kinematics as well as the work piece geometry.[1] The deflections lead to inaccuracies during the handling and assembly process that complicate the compliance with the required tolerances.[2]

A process integrated frequent calibration of the robot kinematic can help to improve the positioning accuracy of the robot. The calibration routine should be quick and repeatable. For a number of kinematics within the same working area a calibration with a calibrated material standard is possible. The material standard displays the same manufacturing and assembly tasks as the later product. Therefore the manufacturing of the standard is time consuming and the method is inflexible and expensive especially for large scale parts. Other methods aim to improve the positioning accuracy of the robot kinematics by the use of Global Reference Systems (GRS). The GRS defines a communication network for the exchange of information such as the position of manipulating machinery and non-manipulating devices of a production system. For manipulating machinery, large-volume measurement systems [3,4] such as indoor-GPS (iGPS) and laser trackers are suitable [5,6,7]. For non-manipulating devices, deviations are detected by local sensors e.g. machine-vision systems, light-section or force/ torque sensors. They are capable of measuring local features and require

S. Ratchev (Ed.): IPAS 2014, IFIP AICT 435, pp. 79–86, 2014.

additional data from other GRS- integrated systems. This information is necessary to transform the 3D CAD data into the reference system and aligned the real work piece to final geometry. The integration of the measurement technologies is expensive and the measurement uncertainty is approximated >100µm without the influence of very harsh environmental conditions. That does not meet the accuracy requirements of many assembly processes.

2   New Measurement Technologies for Maximum Precision

Etalon´s Absolute Multi Line is a new measurement system based on frequency scanning interferometry (FSI) and seems to be capable for a reconfigurable, fast and very precise calibration of a number of robot kinematics in the same working space. The measurement system is able to measure 24 absolute lengths with an uncertainty of $U=0,5+0,5µm/m$ (interferometry uncertainty). The set up consists of 24 optical fibers and a reflector mounted on the Tool Center Point (TCP) of the robot. The laser lines can easily be integrated along typical working paths of the robot. For the calibration the robot will move the reflector into the beam and run a calibration sequence along the laser beam. At each point the measured length between the end of the fiber and the reflector will be compared to the length calculated from the positioning information of the robot. The deviations between the data can be used for an evaluation of the robot´s positioning accuracy and be compared to the accuracy required by the holistic tolerance management. In addition the information shall be used for an online compensation of the robot path. But there is an even more sophisticating solution to enable an online calibration and even control of the robot kinematics. This approach requires the ability to track a target which is mounted on the robot´s TCP by multiple interferometry distance measurements. Therefor the system need the ability to track a target mounted on the TCP of the robot. The mechanical concept is presented in this paper.

3   Virtual Metrology Frame

In the 1960's Brown described the idea of a "virtual metrology frame". Greenleaf illustrated the principle of a selfcalibrating surface measuring machine in 1983 as shown in figure 1.

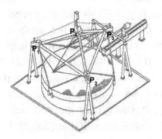

Fig. 1. Selfcalibrating surface measuring machine

The concept is based on the multilateration principle. The fundamental elements of a measuring system based on multilateration are laser-interferometer-based tracking stations, retro reflectors and a mathematical approach to determine the spatial coordinates of the measured points. The "virtual metrology frame" of the present approach is based on the "High-Accuracy CMM" presented in 2000 by Hughes (National Physical Laboratory, UK). The position of the TCP equipped with retro reflecting targets is tracked and measured by fixed measuring stations located at appropriate positions around the working zone of the robot kinematics. As Hughes points out the virtual metrology frame approach will use eight measuring stations or better eight "lines of sight" to achieve a six degree-of-freedom measurement capability.[Design of a High accuracy CMM]

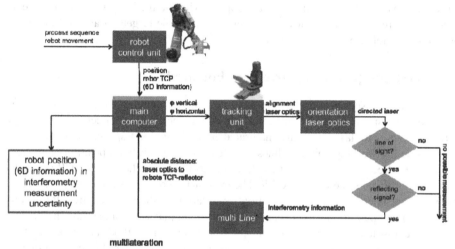

Fig. 2. Description of the tracking system for multilateration

Figure 2 describes the concept for tracking the reflecting elements in detail. The tracking of the targets is based on the operational control data of the robots. The 6D information for position and orientation of the robot´s TCP is sent to the main computer of the multilateral measurement system. Due to the known position of the robot and the tracking units the main computer calculates the vertical and horizontal angle for each rotating stage of the tracking units to align the laser optics with the appropriate reflector which is attached to the robot´s TCP. The aligned laser optics emit the directed laser beam. If there is a line of sight and the reflected signals quality is adequate the interferometry information arrives in the Multi Line. The Multi Line determines the absolute distance between the laser optics and the robot´s TCP- reflector. Information about multiple distances are necessary to use multilateration for calculating the robot´s TCP position. The result is the robot´s TCP position (3D information) in interferometry measurement uncertainty.

The challenges regarding the line of sight and the reflecting signal quality will be addressed in chapter 6 "Boundary condition for absolute interferometry in multi robotic assembly"

This system uses multiple tracking units, which consist of to rotatory stages, to track multiple reflectors. As described the position of the robot´s TCP is known from the robot control unit (figure 2). Each tracking unit can be designed to a more favorable price because the tracking is based on known geometric information and not on complex and active tracking area sensor (in particular PSD). Compared to the laser tracer (each unit has got an own interferometer and a complex active tracking routine) the described concept is using a single multi line system (up to 88 channels) and multiple favorable tracking units which consist of common rotatory stages. By scaling the number of measuring points, this concept is much more favorable than a comparable solution with laser tracers.

An additional advantage because of the absolute interferometry is, that the tracked target can changed by reposition the laser optics with the tracking units to a different robot´s TCP reflector. This allows the control of multiple targets in an assembly cell with a high number of cooperating robots.

4   Test Setup and Improvement Potentials

The test setup consists of 4 robot kinematics, solving cooperating working tasks in the same working space. The TCP of each robot is equipped with a reflector of the measuring system. The laser lines are integrated into the working space. They could be used inside the working space as lines of reference for the robot´s position or movement. The reflector is moved along a laserline to reference the robots TCP position into the coordinate system of the GRS. The information of the absolute interferometry measurement is compared to the position-information of the robot control. Based on the result a compensation routine can be implemented that will compensate the deviation of the absolute positioning. The result is a calibration method that improves the positioning accuracy of the robot kinematics dramatically compared to the accuracies which is reached by the control of multiple robots by Nikon´s iGPS. [2,7]

The tracking unit must provide stable, accurate and continuous changes in the orientation of the laser optics beam in the horizontal and vertical direction. In addition, the pivot point has to be defined as reference point of the optical system. A corresponding prototype of a tracking unit is constructed and assembled out of bought-in components. The objective is to build the kinematics at an affordable price. A detailed description of the development is given in chapter 5.

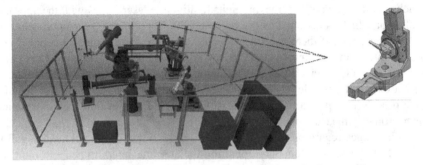

Fig. 3. Test setup for multilateration in cooperative robotic assembly

The calibration methods allows the compensation of system deviations in absolute positioning by implementing a measurement system and a calibration routine into the assembly process. The compensation will improve the positioning accuracy of the robot kinematics and the accuracy of the assembly process. A simulation of the work piece deflections will complement this approach for maximum precision in the assembly of large scale parts. [1]

5   Development of the Tracking Units

To guide the laser optics multiple low price tracking units are necessary. The units must align the collimator to the reflectors, which are attached to the robot´s TCP. The tracking unit´s positioning ability must be accurate enough to hit the reflector area (cateye) with the laser beam over a distance of more than 20 meters.

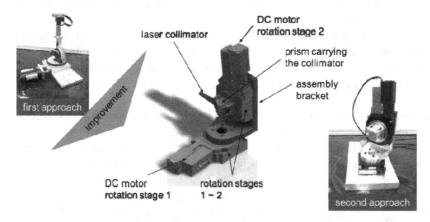

Fig. 4. Tracking unit

Figure 4 shows the developed and improved tracking unit, which consists of two rotating stages (URS 50 BCC Newport) with a minimum incremental motion of 0.001° and a guaranteed absolute accuracy of ±0.04° [8]. Both stages are connected by an assembly bracket and the collimator is carried by an prism.

The improved tracking unit is more robust and can carry heavier loads. The uni- and bi- directional repeatability is improved and the wobble error is minimized. Additionally, the following constructional improvements are realized. The laser collimator can be positioned reliably and reputably in the middle of both rotation axes by the use of the prism and a fixture. The cables of the rotation stage and the optical fiber can be routed through the rotation stage and the ground plate to avoid damage. The upper rotation stage is adjustable in height to investigate and minimize the influence the wobble error of the stages.

In the future piezo-electric driven rotary stage or a goniometer could raise the accuracy of the tracking unit if necessary.

6   Boundary Conditions for Absolute Interferometry in Multi Robotic Assembly

Line of sight visibility and a high signal quality of the reflected laser beam are required to take full advantage of the interferometry system´s accuracy. The complexity of assembly processes and the use of multiple robots result in spatial constraints that increase difficulty to fulfill these requirements.

There are strong dependencies between all (moving) elements in the working area. (e.g. work pieces, fixtures, tooling, grippers, safety fences, robots, etc...) All these elements are obstacles and must be taken into account when planning the assembly process. Also the multiple number of reflectors to determine a position and distance information performing multilateration are challenging.

Additionally the tracking units have got moving constraints and the opening of the reflectors (cat eye) are limited to approx. ±30°. All these boundary conditions limit the scope and solution space for positioning the tracking units and providing a direct line of sight to the reflector.

To handle the described challenges and gain all advantages of interferometry measurement a strong simulation of the metrology infrastructure and the assembly process is needed. All boundary conditions like the robot movements, characteristics of the tracking units and reflectors which are attached to the robot´s TCP must be considered.

To perform this task a simulation is needed to cope with all influencing factors and the domination of complex assembly systems. Figure 5 shows such a metrology infrastructure for multi robotic cooperative assembly which allows the optimization of complex systems including the measurement system in an early planning phase.

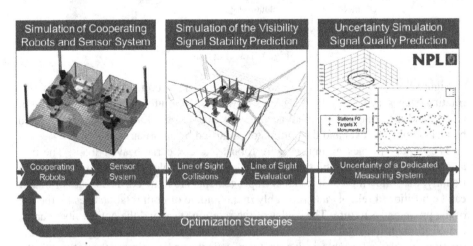

Fig. 5. Metrology infrastructure for multi robotic cooperative assembly

The simulation was developed to examine the boundary conditions of Nikon´s iGPS but can be easily adapted to the requirements of the absolute interferometry measurement. This concept includes not only the line of sight evaluation between the sending and receiving elements but also a model of the measurement systems uncertainty can be implemented to predict the quality of the measurement as well. [9]

The information regarding the line of sight and the predicted measurement uncertainty are used to optimize the assembly system performance for optimum visibility and measurements uncertainty.

7   Optimization Strategy for the Setup of the Assembly Process

If the simulation (presented in figure 5) identifies insufficient number of lines of sights to perform the multilateration measurement, setup optimization can be done by repositioning of the tracking units. Therefore the position of each tracking unit is weighted by a criteria and the relevance of the assembly process is considered.

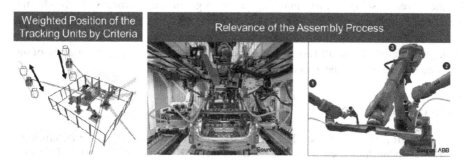

Fig. 6. Approach for the assessment for the optimization of the system´s setup

Figure 6 shows the basis approach for the assessment for the optimization of the system´s setup. The first idea is to weight each position of the tracking unit by a criteria. The criteria evaluate the number of line of sights between the tracking unit position and all reflecting elements during a complete assembly process. The positions with a low criteria value must be optimized to a position with higher value of the criteria. (figure 6, left side)

The relevance of the assembly process is illustrated by two welding operations. (figure 6, right side) The robot paths are colored with respect to their relevance for the process. During the robot movement to the start of the work pieces welding seam the relevance is low (green). When the robot is approaching the first welding tip the relevance rises (yellow) and during the welding process the relevance is high (red).

A heuristic approach is moving the lowest rated (with respect to the positions criteria and the process relevance) tracking unit by ± 1m in the X-Y-Z direction inside the simulation. The new position is evaluated and the rating is compared between each position. Comparing the rating of each position shows the next best setup for the tracking units. [10] Both concepts are considered for finding an optimal position for each tracking unit in respect to visibility and process relevance.

8   Conclusion

This paper introduces an approach to integrate an absolute interferometry measurement into multi robotic assembly applications. Brown´s idea of a "virtual metrology frame" and Greenleaf´s principle of a mechanical selfcalibrating surface measuring machine becomes accessible for multi robotic assembly. The 24 Multi-Line channels are positioned and orientated by cost efficient tracking units based on the operational control data and 3D position information of the robots. This enables the tracking of multiple targets by multiple interferometry distance measurements to determine a 3D position information of the robot´s TCP through multilateration. In the future there will be further developments integrating more reflectors to calculate 6D information for the position and orientation of the robot´s TCP.

A strong simulation of the metrology infrastructure incorporates all influences of the multi robotic assembly process and the measuring system. It is possible to optimize the system´s setup in early planning phase to guarantee a strong visibility during the assembly process.

The concept results in a virtual metrology frame which allows to determine 3D position information for multi robotic assembly with the measurement uncertainty of an absolute interferometry measurements system.

References

1. Schmitt, R., Witte, A., Janßen, M., Bertelsmeier, F.: Modeling of Component Deformation in Self-optimizing Assembly Processes. In: ISMTII 2013: The 11th International Symposium on Measurement Technology and Intelligent Instruments - Metrology - Master Global Challenges, Aachen, p. 222 (2013)
2. Demeester, F., et al.: Referenzsysteme für wandlungsfähige Produktion. In: Brecher, C. (ed.) Wettbewerbsfaktor Produktionstechnik - Aachener Perspektiven, pp. 449–477. Shaker, Aachen (2011)
3. Estler, W.: Large-Scale Metrology - An update. CIRP Annals - Manufacturing Technology 51(2), 587–609 (2002)
4. Puttock, J.M.: Large-Scale Metrology. Ann. of CIRP 27(1), 351–356 (1978)
5. Schmitt, R.: Indoor-GPS based robots as a key technology for versatile production. In: Int. Symposium on Robotics, pp. 199–205. ISR, München (2010)
6. Schmitt, R.: Global referencing systems and their contribution to a versatile production. In: Moreira, A. (ed.) Proceedings of the 2011 IPIN Conference, Universidade do Minho, Guimaraes (2011)
7. Norman, A., et al.: Validation of iGPS as an external measurement system for cooperative robot positioning. The International Journal of Advanced Manufacturing (2012)
8. Newport, http://www.newport.com, product: URS 50 BCC
9. Forbes, A., Schmitt, R., Quinders, S.: Metrology Infrastructure for Multi-Robotic Cooperative Assembly. In: The 11th International Symposium on Measurement Technology and Intelligent Instruments ISMTII, Aachen (2013)
10. Schmitt, R., Quinders, S.: Validation and evaluation of iGPS configurations - description of a tool simulating the lines of sight. In: The Proceedings of the 11th International Symposium on Measurement and Quality Control (ISMQC), Cracow, Poland, September 11-13 (2013)

Application of Deep Belief Networks for Precision Mechanism Quality Inspection

Jianwen Sun[1,2,*], Alexander Steinecker[1], and Philipp Glocker[1]

[1] Microassembly & Robotics, Centre Suisse d'Electronique et de Microtechnique S.A.,
Switzerland
[2] Institute of Neuroinformatics, University / ETH Zurich, Switzerland
{jianwen.sun,alexander.steinecker,philipp.glocker}@csem.ch

Abstract. Precision mechanism is widely used for various industry applications. Quality inspection for precision mechanism is essential for manufacturers to assure the product leaving factory with expected quality. In this paper, we propose a novel automated fault detection method, named Tilear, based on a Deep Belief Network (DBN) auto-encoder. DBN is a probabilistic generative model, composed by stacked Restricted Boltzmann Machines. With its RBM-layer-wise training methods, DBN can perform fast inference and extract high level feature of the inputs. By unfolding the stacked RBMs symmetrically, a DBN auto-encoder is constructed to reconstruct the inputs as closely as possible. Based on the DBN auto-encoder, Tilear is structured in two parts: training and decision-making. During training, Tilear is trained with the signals only from good samples, which enables the trained DBN auto-encoder only know how to reconstruct signals of good samples. In the decision-making part, comparing the recorded signal from test sample and the Tilear reconstructed signal, allows to measure how well a recording from a test sample matches the DBN auto-encoder model learned from good samples. A reliable decision could be made. We perform experiments on two different precision mechanisms: precision electromotors and greasing control units. The feasibility of Tilear was demonstrated first. Additionally, performance of Tilear on the acquired electromotor dataset was compared with the state-of-the-art machine learning based fault detection technique, support vector machine (SVM). First result indicates that Tilear excels the SVM in terms of the Area Under the Curve (AUC) obtained from the Receiver Operating Characteristics (ROC) curve plot: 0.960 achieved by Tilear, while 0.941 by SVM.

1 Introduction

Precision mechanism is widely used for various industry applications, such as precision electromotor for industrial automation systems, greasing control units for microsystems, and so on. Quality inspection for precision mechanism is essential for manufacturers to assure the product leaving factory with expected quality.

---

* Corresponding author.

S. Ratchev (Ed.): IPAS 2014, IFIP AICT 435, pp. 87–93, 2014.

Normally, quality inspection at the manufacturer side is performed by trained experts with different methods. Traditionally, it is accomplished by the experts listening to the sound emitted by the product under different conditions. This subjective assessment is expert individual dependent, and can be influenced by several factors. That brings in the variability in the quality inspection.

Certain algorithms and techniques have been developed to overcome the disadvantages introduced by traditional subjective assessment. These techniques roughly can fall into three categories: signal analysis based methods (SAMs), dynamic model based methods (DMMs), and knowledge based methods (KMs) [1]. With SAMs, experts directly analyze the characteristics of measured signal by performing certain time-frequency transforms, like Fast Fourier Transform (FFT). It is widely used by the industry, but it is always necessary to find a best signal feature before starting the threshold comparison. For some barely seen defect types, it takes a long time to select features. As for the DMMs, an accurate dynamic model for each specific mechanism model is required before performing the quality inspection. KMs have been widely studied recently with the development of machine learning algorithms. Most of these readily available techniques are on the basis of discriminative learning models. A certain amount of fault samples are required to perform the fault type classification [2,3]. However, in practical applications, it is extremely difficult to get fault samples in abundance. What makes matters worse is that even a single type of defect typically has many different sensory manifestations.

Alternatively, we treat the fault detection problem as an anomaly detection problem, to overcome the scarcity of defective samples in the production line. The core of anomaly detection is to recognize the inputs that differ from those under normal conditions. Different anomaly detection techniques have been proposed [4], such as classification based anomaly detection techniques or statistical anomaly detection techniques. J. McBain et al [5] applied the boundary based method, namely Support Vector Data Descriptor (SVDD), with the high dimension signal features extracted by autoregressive model for motor fault detection. They tried to maximize the distance between the average distance of normal class and the average distance of the defective class. B. Zheng et al [6] proposed to use different discriminative classification methods with extracted vibration signal features for bearing anomaly detection. These proposed methods are still discriminative based, which means the construction of the anomaly detector still needs the presence of defective samples, even if not a big number. Also, they usually train with carefully selected features, the choice of which may greatly influence the anomaly detector's performance. Rather than constructing the anomaly detector based on discriminative classifiers, Deep Belief Network (DBN) is selected since it is a generative model which has strong ability to perform fast inference and to learn features unsupervisedly [7-9]. Firstly proposed by Hinton in 2006, DBN has attracted great attention from both academia and industry, and shown promising future in many tasks, such as real time speech translation and image recognition. To our knowledge, our project is the first time that deep belief networks is applied for machinery quality inspection.

The objective of this work is to develop a new automated fault detection system for precision mechanism inspection either using acoustic signals or vibration signals.

Treating the fault detection as an anomaly detection problem, this system is based on a Deep Belief Network (DBN) auto-encoder. It learns the sensory signals only from good samples, and makes decisions for test samples with the trained network.

2   Methods and Results

2.1   Theory Basis: Deep Belief Networks

DBN is a probabilistic generative model, which employs a hierarchical structure constructed by stacking Restricted Boltzmann Machines (RBMs) [8,10]. The RBM is a two layer neural network modeling the joint distribution of its inputs and outputs. To construct a DBN, a number of RBMs are stacked on top of each other. The hidden layers of lower level RBMs are the visible layers of the adjacent higher level RBMs. A greedy layer-wise training algorithm is applied to train the DBN, which is actually training the RBMs individually under the contrastive divergence rule [7]. Trained in this way, the DBN can perform a fast inference and extract high level representations, or features, of the input data. Thorough descriptions of DBNs' mathematical and technical details are available elsewhere [8,10].

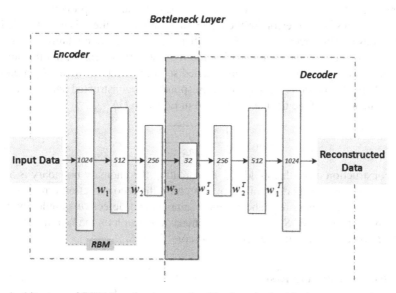

Fig. 1. Architecture of DBN based auto-encoder. Numbers in the blocks represent the number of nodes in each layer. Node number in the bottom layer represents the sampling points from the input data. Node number of the rest layers in the encoder represents the number of extracted high-order features for their respective input data. The number of nodes and layers are only examples. It is not required to have the same numbers in the experiments, or to be 2^n.

By unfolding the stacked n RBMs, an auto-encoder composed by $(2n - 1)$ RBMs is constructed. This $(2n - 1)$ directed auto-encoder can be fine-tuned with backpropagation [9]. As shown in Figure 1, the first n RBMs act as an encoder. High-level features of the input data are extracted by this encoder and stored at the hidden layer of the top RBM. The last n RBMs, including the top RBM of the encoder, form a decoder. This decoder reconstructs the input data with the extracted high-level features stored in the top RBM of the encoder. Generally speaking, a DBN based auto-encoder is to reconstruct training data as closely as possible.

2.2   Proposed Method: *Tilear*

Taking the advantage of DBN auto-encoder's capability to reconstruct the input data as closely as possible, we propose an anomaly detection model, named *Tilear* which learns the input data tile by tile for the purpose of performing fast inference.

Tilear has two functions: "Teacher" for the training phase, and "Tester" for the decision making phase. The biggest difference between *Tilear* and other teacher/tester systems is that during the training phase (teacher), only input data from good samples will be learned by the auto-encoder of *Tilear*, while other systems usually need the presence of defective samples for teacher training. In the training phase of *Tilear*, small anomalies in the "good" data are tolerable variances. The scarcity of anomalies prevents the DBN from learning and reconstructing those. This property results in an additional reconstruction error for the data containing anomalies. Therefore, the higher the reconstruction error, the more anomalies the data sample contains. An anomaly detector thus can be made by comparing the reconstruction error with a threshold. In *Tilear*, the reconstruction error S_i, also named score, is the Root Mean Squared Error (RMSE) between the input data I_i and corresponding reconstructed data R_i, averaged over n dimensions of the data, as expressed in Equation (1).

$$S_i = \sqrt{\frac{\sum_{j=1}^{n}(R_{ij} - I_{ij})^2}{n}} \tag{1}$$

The reconstruction error threshold S_{th} demarcating the anomaly boundary is another model parameter. This is determined by searching the reconstruction error space of a validation dataset containing labeled good samples and defective samples with anomalies. With the selected S_{th}, "Tester" can make a decision on the health status of test sample T_i by comparing its reconstruction score S_i to S_{th}.

2.3   Experiments and Results

Experiments on two different precision mechanisms were accomplished: precision electromotors with a 2 stage planetary gearbox, and greasing control units.

For the precision electromotor, vibration signals were acquired from 36 samples including 21 good samples and 15 defective samples with missing gears on different stages. Cepstrograms of these signals were used as the input for *Tilear*. The distribution of the reconstruction errors is shown in Figure 2.

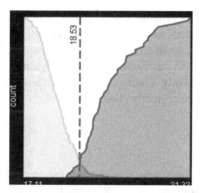

Fig. 2. Cumulative distribution of the reconstruction errors of the precision electromotor dataset. The cumulative distribution of good samples is marked with green, while that of bad samples with red, which is flipped vertically to help examine the overlap between the two distributions. The less overlapped they are, the better performance *Tilear* has. No overlap means the detector can always make the right decision. The threshold selected by *Tilear* for decision making is shown as a red dashed line along with its actual value.

From the above figure, it is observed that most of the defective electromotors with missing gears can be detected. Although there were few samples misclassified with the self-selected threshold, it is possible to filter out all defective samples with the price of some false negative samples.

In order to evaluate *Tilear*'s performance, Area under Curve (AUC) obtained from the Receiver Operating Characteristics (ROC) curve plot is employed, due to the imbalanced class distribution in the electromotor dataset [11]. SVM as a state-of-the-art machine learning algorithm was used as a comparison benchmark. LIBSVM [12] was used to construct the SVM fault detector. Cepstrograms of the vibration signals were used as the input data for both *Tilear* and SVM. The comparison result is shown in table.1. It is observed that the AUC of *Tilear* is higher than that of SVM. This indicates that *Tilear* had a better performance over SVM on this dataset, which has to be verified with further study. It is also worth pointing out that *Tilear* was faster for training compared to LIBSVM. The training time for *Tilear* was approximately 40 minutes, while LIBSVM always took at least several hours. In some sense, it is unfair to compare the training time here, since *Tilear* is developed to use Graphics Processing Unit (GPU) to achieve fast computation speed while LIBSVM not. The comparison of computational speed between *Tilear* and SVM on GPU platform is to be investigated in the future.

Table 1. Comparison of AUC between *Tilear* and SVM

	TILEAR	SVM
AUC	0.960	0.941

As for the greasing control units, acoustic signals were acquired from 47 samples consisting of 24 greased ones, which were considered as good samples, and 23 non-greased ones, which were considered as defective samples. Spectrograms were used as the input of *Tilear*. AUC was again used as the evaluation metric. AUC for these control units was 0.995. Only 1 out of 23 non-greased samples was misclassified, while all greased ones were correctly classified.

3   Conclusion

Tilear, a new automated fault detection method for precision mechanism quality inspection, which firstly uses the Deep Belief Networks (DBN) based auto-encoder, was proposed. *Tilear* is trained to reconstruct the data only from good samples as closely as possible. By comparing the reconstruction errors, a decision can be made. The feasibility of fault detection using *Tilear* is verified with two different kinds of precision mechanisms. It is shown that *Tilear* has comparable performance with the state-of-art technique, Support Vector Machine, using the Area under the Curve as the performance evaluation metric. It is believed that DBN not only can be used for fault detection, but also has the potential in the fault classification area, on condition that enough defective samples are collected for training.

References

1. Liu, X., Zhang, H., Liu, J., Yang, J.: Fault Detection and Diagnosis of Permanent-Magnet DC Motor Based on Parameter Estimation and Neural Network. IEEE Transactions on Industrial Electronics 47(5), 1021–1030 (2000)
2. Yang, J., Zhang, Y., Zhu, Y.: Intelligent fault diagnosis of rolling element bearing based on SVMs and fractal dimension. Mechanical Systems and Signal Processing 21(5), 2012–2024 (2007)
3. Wu, S.D., Wu, P.H., Wu, C.W., Ding, J.J., Wang, C.C.: Bearing Fault Diagnosis Based on Multiscale Permutation Entropy and Support Vector Machine. Entropy 14(12), 1343–1356 (2012)
4. Chandola, V., Banerjee, A., Kumar, V.: Anomaly detection. ACM Computing Surveys 41(3), 1–58 (2009)
5. McBain, J., Timusk, M.: Feature extraction for novelty detection as applied to fault detection in machinery. Pattern Recognition Letters 32(7), 1054–1061 (2011)
6. Zhang, B., Georgoulas, G., Orchard, M., Saxena, A., Brown, D., Vachtsevanos, G., Liang, S.: Rolling element bearing feature extraction and anomaly detection based on vibration monitoring. In: IEEE 16th Mediterranean Conference on Control and Automation, pp. 1792–1797 (2008)
7. Hinton, G.E.: Training products of experts by minimizing contrastive divergence. Neural Computation 14(8), 1771–1800 (2002)
8. Hinton, G., Osindero, S., Teh, Y.: A fast learning algorithm for deep belief nets. Neural Computation 1554, 1527–1554 (2006)
9. Hinton, G., Salakhutdinov, R.: Reducing the dimensionality of data with neural networks. In: Science, vol. 313, pp. 504–507 (2006)

10. Bengio, Y.: Learning Deep Architectures for AI. Foundations and Trends® in Machine Learning 2(1), 1–127 (2009)
11. Chawla, N.V.: Data mining for imbalanced datasets: An overview. In: Data Mining and Knowledge Discovery Handbook, pp. 853–867. Springer, US (2005)
12. Chang, C.C., Lin, C.J.: Libsvm: A Library for Support Vector Machines. ACM Transactions on Intelligent Systems and Technology 2(3), 1–27 (2011)

Visual Quality Inspection and Fine Anomalies: Methods and Application

Simon-Frédéric Désage[1], Gilles Pitard[1], Maurice Pillet[1], Hugues Favrelière[1], Fabrice Frelin[1], Serge Samper[1,2], Gaëtan Le Goïc[1,3], Laurent Gwinner[4], and Pierre Jochum[4]

[1] SYMME, Laboratoire des Systèmes et Matériaux pour la Mécatronique, Université de Savoie, Annecy, France
`{simon-frederic.desage,gilles.pitard,maurice.pillet,`
`hugues.favreliere,fabrice.frelin,serge.samper,`
`gaetan.legoic}@univ-savoie.fr`
[2] LARMAUR ERL CNRS 6274, Laboratoire de Recherche en Mécanique Appliquée de l'Université de Rennes 1, Rennes, France
[3] LE2I, Laboratoire d'Electronique, Informatique et Image, UMR CNRS 6306, Auxerre, France
[4] CETEHOR, Département technique du comité Francéclat, Besançon, France
`{l.gwinner,p.jochum}@cetehor.com`

Abstract. This study develops a surface inspection methodology used to detect complex geometry products and metallic reflective surfaces imperfections. This work is based on combination of three complementary methods: an optical one (structured light information), an algorithmic one (data processing) and a statistical one (parameters processing). *A usual industrial application illustrates this processing.*

Keywords: Surface metrology, Quality Inspection, Computer Vision, Image Processing, Statistic.

1 Introduction

Even nowadays visual quality inspection is a problem. Inspected items have complex geometries, metallic reflective surfaces and high-added values. There have already been studies in industry to establish a methodology for quality inspection, particularly through the PhD works of Anne-Sophie Guerra [1], Nathalie Baudet [2] and Gaëtan Le Goïc [3]. These works helped to highlight the variability of human expert judgment (or controllers) during quality inspection processes [4]. The main objective is to eliminate over and under quality, by improving the controller's working conditions and thus, reduce costs for companies. Our method proposes to use an objective vision system calibrated with Human-based criteria. An example of fine defect automatic detection is presented. The choice of best processing parameters is optimized by design of experiments.

S. Ratchev (Ed.): IPAS 2014, IFIP AICT 435, pp. 94–106, 2014.

2   Background

2.1   Industrial Problem

Following the Anne-Sophie Guerra PhD [1 - p. 85-86] and Nathalie Baudet [2 - p.73], a generalized methodology was established for surface inspection. It consists of three main phases:

1) **Exploration phase:** Anomaly (ies) observation and research - Extraction.

2) **Evaluation phase:** Determination of anomaly criticality.

3) **Decision phase:** Determination of product overall acceptability.

The implementation of this methodology allows visual quality control of products . Through quality inspection, companies want to tighten their evaluation criteria to improve the quality of their products. This is based on the hypothesis that a human control is the best detection system.

That's why, in many companies, quality inspection is still done visually by human operators. This is often justified by the high optical anisotropy of surface appearance of controlled products. This is due either to product complex geometry or its surface texturing that interacts irregularly with light, or a combination of both.

These operators define human visual perception (or sensitivity) like a visual acceptability threshold of product surface appearance. Industrial problem of quality inspection is particularly sensitive to human judgment variability. Indeed, human sensitivity is different for the same person over time and depending upon the mood. Moreover, people have their own sensitivity and experience, which is also a source of variations in human controls. To reduce this variability, we propose a vision system and data processing methods to assist human controller for a more objective detection. We recommend observing an object on a screen in order to have the same view (standard observation) and to agree on the defect importance (criticality).

2.2   Anomalies and Classification

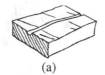

(a)   (b)

Fig. 1. (a) Example of stripe and (b) example of crack. [5]

We have established an easy classification of typical metal-surface products anomalies. Anomalies are defined as differences between studied cases and perfect model. A defect is an unacceptable anomaly. We distinguish two types of light behavior for surface anomalies. Indeed, metal products generally have two types of anomalies:

– The first type is **anomalies of bright appearance** relative to their immediate neighborhood. *Stripes* are a typical example (Figure 1a). These anomalies caused by design, manufacturing or even using.

– The second type is **anomalies of dark appearance** relative to their immediate neighborhood. *Cracks* are a typical example (Figure 1b). Unlike the first type, these anomalies are usually due to faulty manufacturing on small products. [6]

For both, we can see them when there is a high contrast. The difficulty for controllers is to find something that does not exist in a "normal" specimen. They must also be able to do so in such a configuration that replicates any potential orientation that would cause the defect to be unpleasant to the customer, and within a limited timeframe. Generally, either anomaly or its neighborhood blinds us, which is uncomfortable for our eyes.

2.3   Optics Data

Gaëtan le Goïc has applied the generalized methodology [3 - p.5-15] to metallic products. He proposes a visual inspection system (Figure 2), close to a photometric system, which allows a link with human perception [3 - p.126-131]. This system can provide all object views for a hemisphere of illumination, i.e. one view for one lighting incidence, while it is not certain that a human controller has control all of these configurations.

There is therefore an interest in optical systems that allow for an enhanced view of an object's appearance to improve his quality inspection. We made a distinction between vision systems that can observe three-dimensional objects and those that allow understanding object surface by shades of luminance. It is possible to have both in the same device.

For the first category, we can notice all systems as stereoscopy that using multiple shots and giving geometric shape. For the second category, the optical system is generally composed of a dome (hemispherical or almost a complete sphere) on which light sources are regularly distributed, and having a camera at the top of system. The interest of regular arrangement of light sources is to facilitate acquired data interpolation, and so as to understand object surface appearance.

(a)   (b)

Fig. 2. Surface Inspection Support Device, developed at the SYMME Laboratory. [3 –p.130]

The difficulty is to find a compromise between mathematical complexity and data volume, and to have reduced fitting function the closest. In other words, we want to keep photorealistic rendering and optimize memory volume. It should also take into account (spatial and electronic) device resolutions and its optical magnification.

In order to facilitate calculations, models have been used to approximate an interpolation function. We can cite two current types of interpolation:

– Polynomial interpolation (PTM) – [3] [7-9]

– Interpolation by spherical or hemispherical harmonics (SH or HSH) – [10-12]

All of these systems provide an object view in order to reach the first phase of quality inspection: the **"Exploration phase"**.

Moreover, we are working in collaboration with CETEHOR to deal with real cases of watch making with complex geometry, such as watch links. In order to facilitate a reduction in data volume, we used a system that incorporated interpolation. Many different object views can then be reduced to a subset configuration for anomaly detection. The quality criteria for watch links were agreed upon with an expert from CETEHOR.

3   Proposed Method: Structured Detection

3.1   The Image: Essential Data

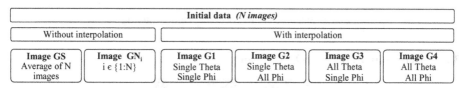

Fig. 3. Diagram of possible image data

We use several "classifications" of images:
- **GS**: That's the image of average rendering of acquired data.
- **GN$_i$**: That's one image from acquired data (number $i \in \{1:N\}$).
  Theta is azimuth and **Phi** is elevation of a hemisphere.
- **G1**: That's the object's rendering from a unique configuration of the three reference points (lighting - object - point of view), as a photo.
- **G2**: That's the object's rendering such as illuminated by a quarter of a vertical continuous circle of leds.
- **G3**: That's the object's rendering such as illuminated by a horizontal ring of leds.
- **G4**: That's the object's rendering such as illuminated by the whole hemisphere.

First, we need to extract or compose an image containing anomaly (ies). In the case of an "a priori" knowledge from product geometry, we can restrict research to a hemispherical space. There is already a system restricted to equivalent of two images G3 for flat objects (Figure 3) [13]. Any global specularity causing glare in the viewing system should be prohibited. In a first optimization, we can work with GS images and we can reduce the work area before filtering. We present a possible diagram (Figure 4) to define a relevant area and to obtain an image that we have labeled RS as reduced image. Its application is illustrated with the original image GS (Figure 5a), resulting image mask GM (Figure 5b) and reduced image RS (Figure 5c). For a complete analysis, the processing must be applied on each image G1.

Fig. 4. Diagram of preprocessing before filtering

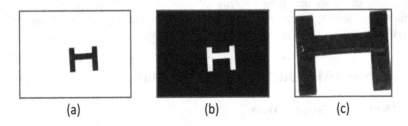

Fig. 5. Example of preprocessing before filtering

3.2   Detection Methods and Constraints

Fig. 6. Example of link of watch. Result of image RS combination and its mask. The viewing side is convex. Two remarkable things: a circle indicates dust and rectangle indicates a stripe. This stripe is evaluated by human controller as a "Strong mark".

Previous works have highlighted several techniques of image processing to detect fine anomalies. These are different combinations of filters such as the **"Top-hat"** [14] or the **"maximum opening"** [15-16], or the **"subtraction of maximum closure of opening by original image"** [17-18]. These methods are illustrated (Figure 7). These filters can easily and quickly become complicated. There are some states of the art for other detection methods such as road defects [6] or fabric defects [19]. For these reasons that is necessary to know characteristics of the object that are sought. These methods are derived from mathematical morphology to extract a local supremum (upper bound of local neighborhood) or a local infimum (lower bound of the local neighborhood). The local neighborhood is defined before processing by a structuring element.

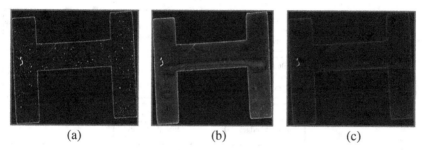

(a)   (b)   (c)

Fig. 7. Example of watch link. (a), (b) and (c) are filtering results of Figure 6. (a) by top-hat (b) by maximum opening and (c) by subtraction of maximum closure of opening by original image.

Structuring element is generally defined by its shape, its length, and, eventually, by direction. In our case, sought anomalies are called fine because they have a linear structure. The structuring element, consistent with these anomalies, will be linear. The mathematical operation then is to browse with the structuring element over the image represented by a matrix of received light data. The result of this is then a processed image. [20]

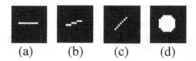

(a)   (b)   (c)   (d)

Fig. 8. Example of structuring element. (a), (b) and (c) are line structuring elements with different orientation (0°, 10°, 45°). (d) is a disk structuring element.

First, a fine anomaly with linear structure has a dominant direction. Depending on the desired filter, the direction of the linear structuring element is either perpendicular or parallel to the anomaly direction. This implies that the direction choice of the linear structuring element is important relative to the anomaly direction. However, our optical system enables taking of pictures structured in relation to the object and the direction of the anomalies, which allows us to constrain the mathematical treatment dedicated to the anomalies detection.

3.3   Proposed Filtering

An anomaly is best revealed when illuminated from a direction perpendicular to its dominant direction. So, the processing can be limited to two directions, parallel and perpendicular to incident direction of light, depending on desired filtering. We can combine filtering operations to obtain information on anomalies as residual information of processing as shown (Figures 9-10). A "twisted" anomaly, consisting of multiple "dominant" directions, can be broken down into several sub-anomalies, and processed as the sum of partial directions.

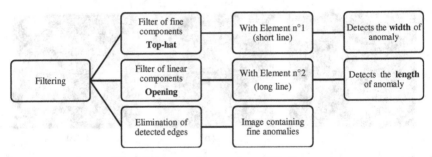

Fig. 9. Diagram of processing of filtering

Firstly, we apply a **Top-hat filter** with structuring element n°1 as a **low-pass filter** in the perpendicular direction to the dominant direction of anomalies. So, we keep elements with thin width in image. For images G1 or G2, we know the incident direction of light, so we can apply the filter in this direction because fine anomalies such as stripes are revealed for this incident direction of light. For images G3, G4 and GS, we must browse in different directions. Sometimes, it is interesting to process only for the four main directions as 0°, 45°, 90° and 135° ±180°.

Secondly, we apply **Opening filter** with structuring element n°2 as **high-pass filter** in parallel direction to the dominant direction of anomalies. So, we keep the elements with high height in the image. When we know incident direction of light, we can apply this filter in the perpendicular direction at the incident direction of light. The direction of this filter has 90° rotated from the Top-hat filter, as it has been hypothesized that fine anomalies have linear structure.

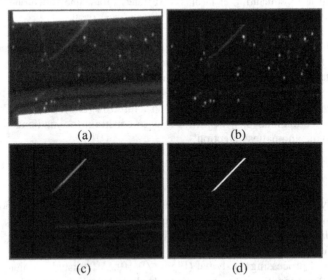

Fig. 10. Example of filtering with a region of interest of watch's link

We can also deduce the size scale for anomalies with a simple formula:

$$S = \left[\frac{FL}{WD} * \frac{SR}{SS} * CS \right] \qquad (1)$$

Where S (pixels) is the structuring element size, WD (mm) is the working distance (camera-object distance), FL (mm) is the camera focal length, SS is the camera sensor size, SR (pixels) is the camera sensor resolution and CS (mm) is the crack appearance size. [19]

3.4   Optimization by Experimental Approach

Like the use of the design of experiments (DOE) method in the field of mechanics to optimize machining parameters [21], we want to use DOE to optimize the choice of the previous parameters, such as the characteristics of the element structuring. The objective is to obtain improved detection and thus facilitate evaluation of surface anomalies [22]. Indeed, this method is suitable when it comes to taking into account a large number of parameters and the relationship between them is not necessarily linear.

The capability to detect an anomaly depends of the structuring element parameters. These parameters are:

-    The length of element 1 and element 2. The length of element 2 depends on length/width ratio of anomaly.
-    The step of element's direction
-    The mask of object

	Factors	Levels				
		1	**2**	**3**	**4**	**5**
A	Mask	Without	Relevant area	Mask complete	Mask without edge (5 pixels)	Mask without edge (10 pixels)
B	Direction step	1°	3°	5°	15°	45°
C	Length of Element 1	5 pixels	10 pixels	15 pixels	20 pixels	25 pixels
D	Length of Element 2	2* Element 1	4*Element 1	6*Element 1	8*Element 1	10*Element 1

Fig. 11. Table settings

Using 5 levels per factor, the **full** factorial design is a 625 line table. We chose to make the fractional design L_{25} 5^6 with 25 trials. (Table 1, see Appendix)

The operation is to optimize two responses to find the best compromise. They are:

1.   The quality of detection

The capability is a measure by comparison between the expert judgment on a parts sample and the quality of the anomaly image. Note the ease of finding the anomaly identified by the expert in a scale from 0 to 10, (Figure 12a)

2. The computation time

Note the ease of process to identify anomaly in a scale of time (seconds) (Figure 12b)

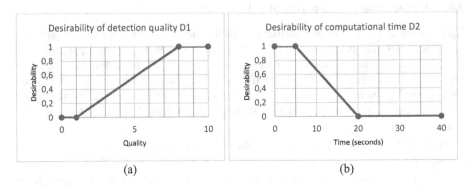

(a)   (b)

Fig. 12. Desirability diagrams of detection, for quality and computational time

The overall desirability is calculated by the geometric mean of the two desirabilities:

$$OD = \sqrt[2]{D1.D2} \qquad (2)$$

The overall desirability is as shown (Figure 15) and is the result of formula (2). The experiment is performed on five links with anomalies identified. The effects graph of quality response and time response are as shown (Figures 13-14). The local optima are surrounded by a red circle. For the parameter of the mask, it was obvious that the reduction of search area (of anomalies) would save computation time. However, the effects graph shows the reduction of the area may lead to wrong detection by generating false positives, such as edge detection.

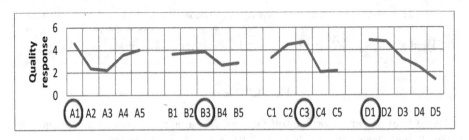

Fig. 13. Graph of the effects of quality response (average of 5 samples)

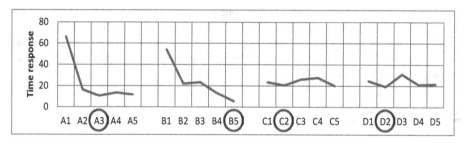

Fig. 14. Graph of the effects of time response (average of 5 samples)

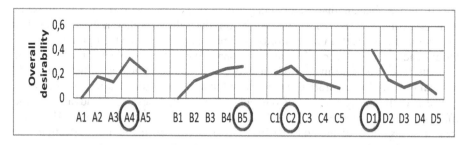

Fig. 15. Diagrams of overall desirability of detection (average of 5 samples)

The optimized parameters are **mask with trimmed edges 5 pixels**, a **step of 45 degrees**, a **length of 10 pixels for element 1** and a **width/length ratio of 2** to define length of element 2. These parameters are applied (Figure 16). The final image (Figure 16b) is used to assess the quality of the product based on the residual information it contains.

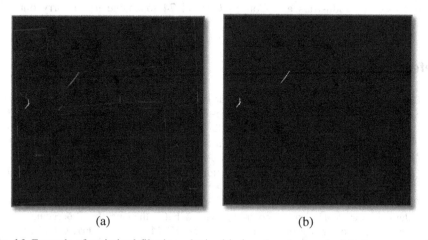

(a)   (b)

Fig. 16. Example of optimized filtering, obtained in less than 4 seconds. (a) shows areas with visual impact on the links without edges and (b) is the result of threshold of these areas above half the maximum intensity.

With formula (1), we have an idea of appearance width of anomaly. The working distance (camera-object distance) is 440 mm, the camera focal length is 440 mm, the camera sensor size is 15.8 mm, the camera sensor resolution is 2136 pixels and the size of structuring element is between 10 and 15 pixels. So the appearance width is between 75 and 110 μm.

4   Conclusion

Following the methodology formalization of quality inspection, we are able to provide a theoretical meta-model and a physical system for visual exploration and detection of surface anomalies, including product reflective metal surface and those of complex geometry. The main objective is to reduce the overall variability of quality inspection of a product. In other words, we are able to determine the optimum optical system and data processing dedicated to controlling types of defects on one type of products. The main information of our publication is a combination of three methods (**optical - algorithmic - statistic**) to allow automatic detection of surface anomalies

The underlying information of our publication is that despite the advance of the effective methods, companies outside research still have little awareness. Here, we present an example of method that may be of interest in industrial production, but each method has been at least 5 years out of industry. The laboratory aims to transfer technological know-how, more or less mature, to improve industrial production. This means that research (academic) centres, technical centres, and companies must communicate with each other more.

Acknowledgments. We also wish to thank CETEHOR and companies have provided us the necessary samples to our research. We thank our partners in MESURA project as well as Arve Industries and Conseil General 74 to enable us to carry out this research by giving us resources.

References

1. Guerra, A.S.: Métrologie sensorielle dans le cadre du contrôle qualité visuel. Université de Savoie (2008), http://hal.archives-ouvertes.fr/tel-00362743/
2. Baudet, N.: Maitrise de la qualité visuelle des produits – Formalisation du processus d'expertise et proposition d'une approche robuste de contrôle visuel humain. Université de Grenoble (2012), http://tel.archives-ouvertes.fr/tel-00807304/
3. Le Goïc, G., Samper, S.: Système de détection d'anomalies d'aspect par la technique PTM (2011), http://hal.archives-ouvertes.fr/hal-00740313/
4. Baudet, N., Pillet, M., Maire, J.L.: Proposition d'une approche méthodologique pour réduire la variabilité dans le contrôle visuel à but esthétique. In: Proceeding of the International Conference on Surface Metrology, ICSM (2012), http://hal.univ-savoie.fr/hal-00740267/
5. ISO-8785 Geometrical Product Specification (GPS) – Surface Imperfection - Terms, definitions and parameters - International Organization for Standardization (1998)

6. Nguyen, T.S., et al.: Etude d'un algorithme de détection de défauts sur des images de chaussées. XXIIe colloque GRETSI (Signal and Image Processing). Dijon (2009), http://documents.irevues.inist.fr/handle/2042/29100

7. Dellepiane, M., et al.: High quality PTM acquisition: reflection transformation imaging for large objects. In: Proceedings of the 7th International Conference on Virtual Reality, Archeology and Intelligent Cultural Heritage (2006), http://dl.acm.org/citation.cfm?id=2384330

8. Duffy, S.: Polynomial texture mapping at roughting linn rock art site. In: Proceeding of the ISPRS Commission V Mid-Term Symposiumm'close range image measurement techniques (2010), http://www.isprs.org/proceedings/XXXVIII/part5/papers/159.pdf

9. Baril, J.: Modèles de representation multi-résolution pour le rendu photo-réaliste de matériaux complexes. Université Sciences et Technologies Bordeaux 1 (2010), http://hal.archives-ouvertes.fr/tel-00525125/

10. Palma, G.: Visual appareance: Reflectance transformation imaging, RTI (2013)

11. Tunwattanapong, B., et al.: Acquiring reflectance and shape from continuous spherical harmonic illumination. ACM Transactions on Graphics (TOG) 32(4) (2013)

12. Elhabian, S.Y., et al.: Towards efficient and compact phenomenological representation of arbitrary bidirectional surface reflectance. In: British Machine Vision Conference, Dundee (2011), http://mecca.louisville.edu/wwwcvip/research/publications/Pub_Pdf/2011/Shireen/Elhabian_BMVC2011.pdf

13. Zamuner, G.: Application of artificial vision to the quality inspection of surfaces of luxury products. Ecole Polytechnique Fédérale de Lausanne (2012)

14. Zheng, H., et al.: Automatic inspection of metallic surface defects using genetic algorithms. Journal of Materials Processing Technology (2002)

15. Morard, V.: Detection de structures fines par traitements d'images et apprentissage statistique: application au contrôle non destructif. Ecole nationale supérieur des Mines de Paris (2012), http://imanalyse.free.fr/publications/Morardi-2012-These.pdf

16. Morard, V., et al.: One-dimensional openings, granulometries and component trees in per pixel. IEEE Journal of Selected Topics in Signal Processing 6(7) (2012), doi:10.1109/JSTSP.2012.2201694

17. Jahanshahi, M.R., et al.: An innovative methodology for detection and quantification of cracks through incorporation of depth perception. Machine Vision and Application 24(2) (2011), doi:10.1007/s00138-011-0394-0

18. Jahanshahi, M.R., Masri, S.F.: A new methodology for non-contact accurate crack width measurement through photogrammetry for automated structural safety evaluation. Smart Materials and Structures 22(3) (2013), doi:10.1088/0964-1726

19. Ngan, H., et al.: Automated fabric defect detection – A review. Image and Vision Computing 29(7) (2011), doi:10.1016/j.imavis.2011.02.002

20. Serra, J.: Morphological filtering: An overview. Signal Processing 38(1) (1994), doi:10.1016/0165-1684(94)90052-3

21. Pillet, M.: Les plans d'expériences par la method Taguchi (2001), http://hal.archives-ouvertes.fr/hal-00470004/

22. Moon, H., Kim, J.: Intelligent crack detecting algorithm on the concrete crack image using neural network. In: Proceedings of the 28th ISARC (2011)

Appendix

Table 1. The fractional design L_{25} 5^6 with 25 trials

Test n°	N° Factor				Quality (0 - 10) Sample n°					Time (seconds) Sample n°				
	A	B	C	D	1	2	3	4	5	1	2	3	4	5
1	1	1	1	1	6	6	6	6	6	102,8	104	101	113	109
2	1	2	2	2	8	8	8	8	8	64,68	64	63	69,2	65,4
3	1	3	3	3	8	8	8	8	5	81,98	86	81	94,9	89
4	1	4	4	4	0	0	1	1	1	50,13	51	49	57,7	53,3
5	1	5	5	5	1	1	1	1	1	18,33	19	18	21,9	19,8
6	2	1	4	5	1	1	1	2	2	65,46	66	66	63,5	67,1
7	2	2	5	1	2	1	3	3	2	8,24	9	8,9	8,5	8,63
8	2	3	1	2	2	3	6	6	2	3,22	3,5	3,4	3,4	3,44
9	2	4	2	3	2	2	7	2	2	2,54	2,7	2,7	2,59	2,68
10	2	5	3	4	1	1	1	2	2	1,84	1,9	1,9	1,87	2
11	3	1	2	4	2	2	1	2	2	25,13	28	27	25,7	26,2
12	3	2	3	5	1	1	2	3	2	15,64	17	17	15,6	16,3
13	3	3	4	1	2	1	6	6	2	4,68	5	5	4,76	5,01
14	3	4	5	2	2	1	2	3	2	3,32	3,7	3,5	3,37	3,43
15	3	5	1	3	3	2	1	2	2	1,58	1,7	1,6	1,61	1,66
16	4	1	5	3	1	1	2	3	2	47	55	50	47,7	49,4
17	4	2	1	4	0	8	8	0	8	5,99	6,8	6,4	6,02	6,17
18	4	3	2	5	1	1	1	8	2	7,02	7,8	7,3	7,08	7,22
19	4	4	3	1	1	9	8	9	2	2,22	2,5	2,3	2,25	2,3
20	4	5	4	2	1	1	7	3	2	1,76	2	1,8	1,8	1,83
21	5	1	3	2	8	8	8	9	3	21,8	25	23	22,6	23,3
22	5	2	4	3	1	1	1	4	3	13,26	15	14	13,3	14,3
23	5	3	5	4	1	1	1	4	3	12,82	17	14	13,4	14,1
24	5	4	5	4	1	1	1	4	2	4,8	5,7	5,1	5,07	5,04
25	5	5	2	1	6	6	8	8	7	1,57	1,8	1,6	1,6	1,65

Control Methods in Microspheres Precision Assembly for Colloidal Lithography

Olivier Delléa[1], Olga Shavdina[1], Pascal Fugier[1], Philippe Coronel[1],
Emmanuel Ollier[1], and Simon-Frédéric Désage[2]

[1] L2CE, Laboratoire des Composants pour le Conversion de l'Energie, CEA/LITEN,
Laboratoire d'Innovation pour les Technologies des Energies Nouvelles et des nanomatériaux,
Grenoble, France
[2] SYMME, Laboratoire des Systèmes et Matériaux pour la Mécatronique, Université de Savoie,
Annecy, France

Abstract. Colloidal lithography based on the assembly of microspheres is a powerful tool for the creation of a large variety of two dimensional micro or nanostructures patterned. However very few studies examine the control, qualification and quantification of the ordering of the particles once deposited on the substrate. We have developed two unique methods working at microscopic and macroscopic scales, respectively called Microfixe® and Macrofixe®, for the analysis of grain morphology in the case of hexagonal closed packed (HCP) monolayers of spherical microparticles. The processing of the images taken at microscopic scale uses Delaunay triangulation and histograms of lengths and orientations of Delaunay triangles sides. At the macroscopic scale, six camera images are required of the sample illuminated under six different incidence angles. Image treatment consists in the comparison of the six images and eventually subdivision of these images to sharpen the analysis. At the end, the two softwares constitute artificial images of particle deposit, giving at microscopic and macroscopic scales significant information about grain size, grain morphology, orientation distributions, defaults (voids, stacking)… With these two new control methods, colloidal lithography is emerging as an industrial process.

Keywords: Metrology, Quality Control, Computer Vision, Image Processing. Quad tree, Microfixe®, Macrofixe®, Colloidal lithography, micro/nanotechnologies, thin films, tribology, BooStream®.

1 Introduction

Colloidal lithography based on the precision assembly of microspheres is an easy, inexpensive, efficient, and flexible manufacturing approach [1]. It enables the creation of a large variety of two dimensional micro or nanostructures patterned with a high degree of control and reproducibility [2]. Many application areas are concerned by this technology such as mechanics [3], sensors, photonics, surface wetting [4], biological and chemical sensing. One can find many studies aiming at the

S. Ratchev (Ed.): IPAS 2014, IFIP AICT 435, pp. 107–117, 2014.
© IFIP International Federation for Information Processing 2014

self-assembly control or ordering of particles acting on the physical processes such as activation of particle surfaces, solvents, surface pressure... but very few of them examine the control, qualification and quantification of the ordering of the particles once deposited on the substrate.

In the case of silica microspheres, the inherent 2D periodicity constituted by a hexagonal-close-packed (HCP) type ordering (Figure 1) gives rise to a rich variety of interesting optical properties related with photonic crystals.

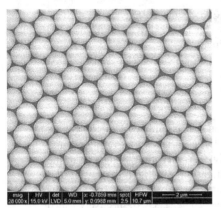

Fig. 1. Hexagonal-close-packed (HCP) monolayer of silica microspheres ∅1.1 μm

Fig. 2. Color response of an HCP monolayer of silica microbeads ∅1.1 μm on diamond like carbon substrate 2x2 cm²

As a result of photonic band gap properties, the 2D-ordered microstructure creates intensive structural colors [5] [6] as shown in Figure 2. In fact, due to process parameters (particle size distribution, surface pressure, withdrawing speed...), the particle deposits are composed of "grains" [7] separated by "fractures". Each grain is composed of hundreds or thousands of particles forming a hexagonal mesh with a specific planar orientation (Figure 3).

Fig. 3. Microscope image (x100) of HCP silica microspheres (∅ 1.1μm) displaying grains separated by fracture lines

In order to master and improve the process, it is essential to control/qualify/ quantify in a rigorous way the grain sizes, grain orientations, fractures, etc. of the particle deposits.

The following paragraphs briefly introduce the precision assembly technique of microspheres, developed by CEA LITEN called BooStream®, to create an HCP monolayer. The fields of application are numerous and we present succinctly its using in the field of tribology. The two analysis methods developed by CEA LITEN and based on image processing and vision are then detailed. The first is software, named Microfixe®, which addresses the analysis of images taken at microscopic scale with optical microscope or scanning electron microscope. The second method, called Macrofixe®, analyses visual rendering of particle deposit (iridescence) by processing a sequence of camera images taken with various lighting conditions.

2   HCP Particles Monolayer Assembly

2.1   Presentation

Particle self-assembly methods at the air/liquid interface are mainly based on the minimization of free energy: the formation of the monolayer is mainly due to the energy level reduction at the air/liquid interface.

Particles then order themselves [8] under different phenomena, such as dipole moments, partial positive charges and hydrophobicity attraction. Common techniques presented in the scientific literature are the Langmuir-Blodgett method [9], vortical method [10] and floating-transferring method [11].

The Boostream® process has been developed on the basis of previous work for the production of active or passive components in the field of energy. In its basic functions, this process uses a moving liquid on which particles are dispensed (Figure 4). Brought by hydrodynamic forces to a transfer area using a slope, the particles are arranged as a film on the liquid. The substrate set previously in contact with the compact film of particles through a capillary bridge is then withdrawn to transfer the film.

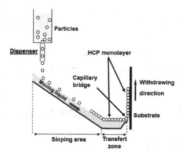

Fig. 4. The BooStream® process (basic configuration)

In this basic configuration, the main advantage of this process is the ability to deal with online substrates of large surface, 2D or 3D, rigid or flexible, for industrial applications.

2.2   Example of Application: Tribology

For several years, surface texturing has been introduced to improve tribological properties of lubricated surfaces [12]. The presence of artificially created micro-dimples on a frictional surface can generate substantial reduction in friction and surface damage when compared with non-textured surfaces. This improvement is attributed to several physical mechanisms like wear debris entrapment, the creation of local increase of lubricant supply by fluid reservoirs and also the increase of load-carrying capacity by a hydrodynamic effect [13].

By combining colloidal lithography and oxygen plasma etching, textured diamond-like amorphous carbon (DLC) films can be produced. The process consists of depositing the monolayer of spherical particles on the DLC film and etching this layer with plasma through the hard mask of silica spheres, then to remove the particles from the substrate by immersion in water with ultrasonic waves. The DLC surface obtained by this method shows micro-pillar networks with dimensions that modulate the surface properties of this material (Figure 5).

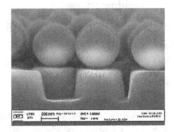

Fig. 5. Micro-pillar networks in DLC film with, on top, HCP silica particle deposit

By this technology, a reduction of 50 % of the friction coefficient has been demonstrated in lubricated conditions [14].

At present, only a few mechanical components used in industrial applications are textured because each application requires an accurate correlation of the texture (depth, density, shape…) with the specific mechanical parameters (geometry, dimensions, load, speed, lubricant…).

This example demonstrates the importance of developing the following control processes described in this paper to apply this technology to industry.

3   Control Method at Microscopique Scale : Microfixe®

3.1   Image Processing Methodology

Image processing is the main tool for the qualification/quantification of the number, size and orientation of the grains. A preliminary step is the extraction of the centre of each particle. Then a Delaunay triangulation [15] is done on this set of points, linking each point with its six (or less) neighbours. It will be useful in the sequel to consider that each link is materialized by a line segment. The orientation of these line segments (with a 60 ° periodicity) is then fundamental.

A simple representation is obtained with the information given by the lengths and orientations of triangle histograms (Figure 6).

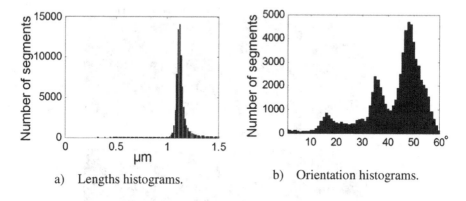

a)   Lengths histograms.   b)   Orientation histograms.

Fig. 6. Analysis of Figure 3, histograms of Delaunay triangles segments

These histograms gather essential information for the determination of quality of the ordered deposits. If the histogram of the triangles' segments lengths distribution is centered on the value corresponding to the nominal particles average diameter, we can deduce that the deposit is compact with few fractures. Moreover, if the histogram of the sides of the triangles' orientations has a single mode (single peak) between 0 ° and 60 °, it means that the deposit of particles is composed of a single grain. In another way, if the triangles' lengths histogram has more than one peak, it means that the deposit of particles is fractured and composed of more than a single grain. In the example in Figure 6, the deposit has several grains with three main directions related to the three peaks on the histogram.

Furthermore, the Delaunay triangulation can be used for the calculation of the relative areas occupied by the different grains on the image giving an indication of the arrangement quality.

3.2   Results

Microfixe® software allows an accurate analysis of the quality of an HCP particles monolayer. Figure 7 shows two cases. The conditions of the deposition process are different in each case. In this example, two images taken with an optical microscope are used. The Delaunay triangulation of these two images is coloured in order to provide an enhancement of the deposit structure, assisting the monitoring of the process.

Assuming that the maximization of grain size is the main criterion for the developed process, the calculation results clearly show that process n°2 overwhelms process n°1 with only three grains and two main orientations. The surface ratio between the ordered area, and the total area analyzed is also a thorough quality index, as shown by the table summarizing the results of Microfixe® software analysis.

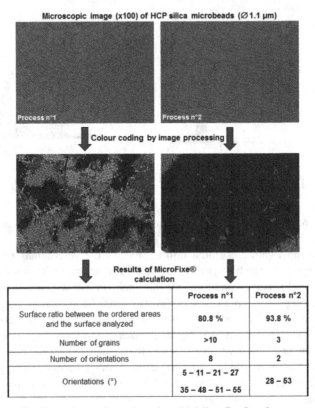

	Process n°1	Process n°2
Surface ratio between the ordered areas and the surface analyzed	80.8 %	93.8 %
Number of grains	>10	3
Number of orientations	8	2
Orientations (°)	5 – 11 – 21 – 27 35 – 48 – 51 – 55	28 – 53

Fig. 7. Analysis of two deposits with Microfixe® software

4   Control Method at Macroscopic Scale: Macrofixe®

Colour is a perceptual attribute created by the human brain in reaction to the stimulus of a certain light signal in a given context. At macroscopic scale, the grains forming part of the deposit have different colours due to the different orientations of the particle structure, which have influence on light diffraction, therefore on the visual appearance. In order to study the correspondence between particle structure and visual rendering, we can observe and analyze deposits with a RGB camera.

4.1   Structural Analysis by Multi-angle Optical Illumination

The diffraction pattern of HCP deposited microspheres for a collimated light has a specific structure displaying symmetry. This pattern is composed of a specular reflection of light (zero order diffraction) surrounded, at equal distance, and distributed every 60 °, by six diffracted beams at the first order of diffraction. The figure below illustrates the diffraction at the first order of a collimated beam having a wavelength λ and incidence angles (θ_i , φ_i) on a screen positioned perpendicular to the propagation direction of the reflected beam (θ_r , φ_r).

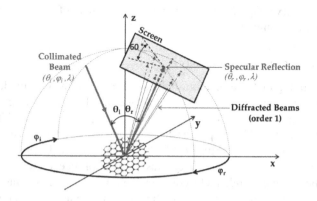

Fig. 8. Illustration of the diffraction phenomenon by HCP structure at zero and first order diffraction on a screen

This angle periodicity indicates that an angular variation of incident light with respect to the sample of 60 ° in the xOy plane (φ_i) is enough to get diffracted towards a fixed point of observation on any HCP structure whatever its orientation in the xOy plane. On this basis, the Macrofixe® control method has been developed using only six illumination and/or observation positions with 10 ° relative angle variations in the xOy plane (φ_i) to obtain the particle structure orientation in all grains.

The six images presented in Figure 9 are a view of the same sample captured from the same point of view, with six relative positions of the light source $(\varphi_i=0, 10$ °, 20 °, 30 °, 40 °, 50 °). This sample is a 10x10 mm² silicon substrate with HCP silica particles of 1.1 μm diameter. The local colour of the sample is related to the orientation of incident light and to the particle orientation. In a given illumination-observation configuration, grains are coloured if they diffract light towards the observation point (camera), otherwise they appear black. Colour is thus a cartographical manner of indicating the local structure orientations of a large sample.

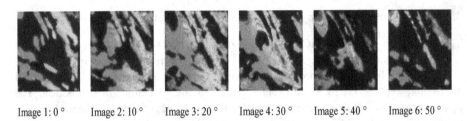

Image 1: 0 °   Image 2: 10 °   Image 3: 20 °   Image 4: 30 °   Image 5: 40 °   Image 6: 50 °

Fig. 9. Images of the same silicon substrate sample (10x10mm²) with a deposit of 1.1μm diameter silica particles for six angle positions (φ_i) of a collimated white light source

Based on these six images, it is possible to extract a map of grains at macroscopic scale by image processing.

4.2   Image Processing Methodology

The six images in Figure 9 are first thresholded according to the method developed by Otsu [16] to create binary images. The percentage of white pixels of each image is then calculated. Figure 10 shows the dependence of the percentage of white pixels on six consecutive images. The amplitude between the min and max curve values is calculated. The greater the amplitude of the curve, the greater the observed structure can be considered as single crystal. In contrast, low amplitude shows that the optical response of the sample is relatively constant regardless of the lighting conditions. Therefore, the observed zone shows no particular structure and can be considered as scattering. Moreover, in the case where the amplitude is high enough, the image corresponding to the maximum of the curve indicates the main direction of diffraction and hence the orientation of the observed structure.

With this approach, it is therefore possible to map the grain structure but also to determine the orientation of the hexagonal structure for each of them.

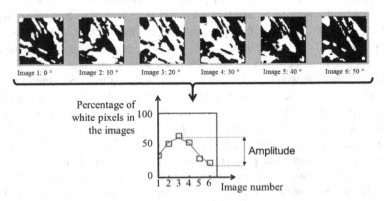

Image 1: 0 °   Image 2: 10 °   Image 3: 20 °   Image 4: 30 °   Image 5: 40 °   Image 6: 50 °

Percentage of white pixels in the images

100

50

0

1 2 3 4 5 6   Image number

Amplitude

Fig. 10. View of binary images from Figure 9 and evolution curve of the white pixels percentage according to the six consecutive images

In accordance with our criteria, in the case where the amplitude is in the range of 90 % to 100 %, the observed area is formed by a single crystal (or near), the system can indicate a high quality of ordering. If the amplitude is between 0 % and 20 %, and the observed area scatters the light, the particles are not organized. If the difference is between 20 % and 90 %, the area cannot be characterized accurately and requires a more detailed analysis. Based on an analysis methodology called QuadTree [ ], each of the six images is then divided into four equal sectors. For each of the four areas an evolution curve of the white pixels number is performed. Similarly, the amplitude of each plotted curve is calculated to qualify characterise each division of the image. In the developed process, the subdivision of the image continues until a stop criterion is reached: ordered area, diffusing area, image size too small (only 1 pixel) etc.

4.3   Results

Based on the protocol described above, the Macrofixe® process is able to reconstitute an image of squares indicating the morphology of the grains and for each, its relative orientation in the plane of the hexagonal lattice. Figure 11 shows for example the treatment of images presented in Figures 9 and 10. The latter shows a large number of grains and many changes of the hexagonal lattice orientations.

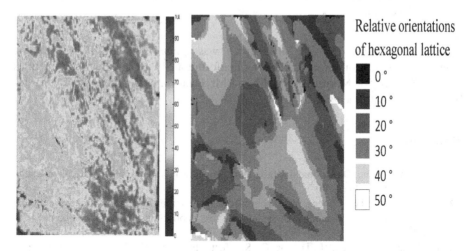

Fig. 11. Resulting image after subdivision of images from Figure 9 following the Macrofixe® method. Grain morphology (left) and hexagonal lattice relative orientations (right).

5   Conclusion and Perspectives

The recent advances made by CEA LITEN in the field of image processing and vision for the control and qualification of hexagonal-close-packed assembly of spherical microparticles have been presented. These control methods, called Microfixe® and Macrofixe®, are able to give significant information respectively at microscopic and macroscopic scales. From SEM or optical microscope images, using Delaunay triangulation, parameters such as surface ratio, number of grains, number of orientations and values of orientation can be extracted. From images taken by camera and thresholded to obtain binary images, the morphology of the grains at the macroscopic scale can be mapped. Useful information (mapping and orientation of the grains) is obtained first by processing six images of a sample illuminated under six different incidence angles separated by 10° in the plane xOy. For each image, the percentage of white pixels is calculated and then compared. Depending on the differences obtained, it is possible to determine whether the observed zone is properly structured, diffusing or not determined. In the case where it is not determined, the six original images are divided into four (for example). The divided parts of the image are then analyzed using the same protocol: calculation of the white pixels percentages, or subdivision if the gap does not allow a definition of the quality of organization. Step by step, an artificial image of the particle deposit is constituted, giving precise information on the morphology of the grains and their orientations.

Microfixe® and Macrofixe® control methods are key points in order to provide an accurate control and enhancement of the deposit structure, providing assistance in the monitoring of the process for rigorous developments in the area of self-assembled materials. Future work will be aimed at developing a real time macroscopic analysis, in order to implement this control method system in an industrial equipment of particles deposition.

Acknowledgements. The present research was funded by the FP7 programme "Flexible Compression Injection Molding Platform for Multi-Scale Surface Structures (IMPRESS). To Mr Martin RANSOM and Mr Jean-Marie Becker (CPE Lyon), for guidance and adjustments.

References

1. Zhang, G., Wang, D.: Colloidal Lithography, The Art of Nanochemical Patterning. Chem. Asian J 4, 236–245 (2009)
2. Ye, X., Qi, L.: Two-dimensionally patterned nanostructures based on monolayer colloidal crystals: Controllable fabrication, assembly, and applications. Nano Today 6, 608–631 (2011)
3. Chouquet, C., Gavillet, J., Ducros, C., Sanchette, F.: Effect of DLC surface texturing on friction and wear during lubricated sliding. Materials Chemistry and Physics 123, 367–371 (2010)
4. Chen, J.-K., Qui, J.-Q., Fan, S.-K., Kuo, S.-W., Ko, F.-H., Chu, C.-W., Chang, F.-C.: Using colloid lithography to fabricate silicon nanopillar arrays on silicon substrates. Journal of Colloid and Interface Science 367, 40–48 (2012)
5. Moon, G.D., Lee, T.I., Kim, B., Chae, G., Kim, J., Kim, S., Myoung, J.-M., Jeong, U.: Assembled Monolayers of Hydrophilic Particles on Water Surfaces. ACS Nano 5(11), 8600–8612 (2011)

6. Retsch, M., Zhou, Z., Rivera, S., Kappl, M., Zhao, X.S., Jonas, U., Li, Q.: Fabrication of Large-Area, Transferable Colloidal Monolayers Utilizing Self-Assembly at the Air/Water Interface, Macromol. Chem. Phys. 210, 230–241 (2009)
7. Hillebrand, R., Muller, F., Schwirn, K., Lee, W., Steinhart, M.: Quantitative analysis of the grain morphology in self-assembled hexagonal lattices. ACS Nano 2(5), 913–920 (2008)
8. Acharya, S., Hill, J.P., Ariga, K.: Soft Langmuir–Blodgett Technique for HardNanomaterials. Adv. Mater 21, 2959–2981 (2009)
9. Bardosova, M., Pemble, M.E., Povey, I.M., Tredgold, R.H.: The Langmuir-Blodgett Approach to Making Colloidal Photonic Crystals from Silica Spheres. Adv. Mater. 22, 3104–3124 (2010)
10. Pan, F., Zhang, J., Cai, C., Wang, T.: Rapid Fabrication of Large-Area Colloidal Crystal Monolayers by a Vortical Surface Method. Langmuir 22(17), 7101–7104 (2006)
11. Zhang, Y.J., Li, W., Chen, K.J.: Application of two-dimensional polystyrene arrays in the fabrication of ordered silicon pillars. Journal of Alloys and Compounds 450, 512–516 (2008)
12. Ninove, F.-P.: Thesis from Ecole centrale de Lyon (France), Texturation de surface par laser femtoseconde en régime élastohydrodynamique et limite, tel-00688051, version 1, ECL 2011-43 (2011)
13. Chouquet, C.: Thesis from l'institut National Polytechnique de Lorraine, Elaboration et caractérisation de revêtements type « Diamond-Like Carbon »déposés par un procédé chimique en phase vapeur assisté par un plasma basse fréquence (2008)
14. CEA LITEN, internal results, contact: olivier.dellea@cea.fr
15. Cocquerez, J.-P., Philipp, S.: Analyse d'images: Filtrage et segmentation, ISBN 2-225-84923-4. (1995) Masson (ed.)
16. Otsu, N.: A threshold Selection Method from Gray-Level Histograms. IEEE Transactions on Sytems, Man, and Cybernetics smc-9 (1979)
17. Finkel, R.A., Bentley, J.L.: Quad Trees A Data Structure for Retrieval on Composite Keys. Acta Informatica 4, 1–9 (1974)

A Multi-Agent System Architecture
for Self-configuration

Nikolas Antzoulatos, Elkin Castro, Daniele Scrimieri, and Svetan Ratchev

Faculty of Engineering, University of Nottingham, Nottingham, United Kingdom
{eaxna6,elkin.castro,daniele.scrimieri,
svetan.ratchev}@nottingham.ac.uk

Abstract. Due to constant globalisation new trends on the market are coming up. One of the trends is the customisation of products for the customer and shorter product life cycles. To overcome the trends industries identified as key element self-reconfigurable production systems. A change to a running system means loss of time, money and manpower. A reconfigurable production system can automatically adapt to changes in terms of changing a machine or a product. The methodology behind is adapted from the office world and is called plug and produce. However, a production system has different requirements which need to be met. Due to a lack of homogeneity of industrial controllers in terms of communication and reconfigurability, as well as the interaction with the end user, the multi-agent technology was identified as a superior communicator. We present a new multilayered multi-agent architecture where the necessary agent types are introduced to fulfil the requirements for plug and produce. One scenario is shown where the architecture is employed to enable plug and produce capabilities and allow the system to adapt itself.

Keywords: Plug and Produce, Multi Agent System, Self*, Robotic Systems.

1 Introduction

Throughout the globalisation, European markets are becoming more dynamic and applying new trends in industries. Therefore, vendors must now deal with more competitors than before. The available products will increase, which in turn will reduce the product life cycles. One way to remain competitive is to supply product variants instead of new technologies, because of the high development costs and time constrains to offer a new product. Another trend to push the selling of products is the customisation of products which is well known from the auto mobile industry. The customer can first choose from a catalogue his favourite equipment and then assemble a personalised car.

The main challenge for European SMEs (small and medium enterprises) is to respond quickly to market changes. A new manufacturing system must cope with it by adjusting and reconfiguring itself with minimum human intervention. New manufacturing processes should be introduced easily to the production line with a minimised downtime. The production should adapt in layout and configuration to different types

S. Ratchev (Ed.): IPAS 2014, IFIP AICT 435, pp. 118–125, 2014.

of products. After a machine failure the production line should recover quickly either by replacing devices or by redirecting the workflow and spreading the tasks to different machines.

The described production issues can be addressed by using a technique called plug and produce [1-3] (otherwise known as plug and work [4-6] or plug and operate [7-9]). The concept enables to plug and unplug a device (like a robot, machine, sensor etc.) with less effort in terms of reconfiguration, reprogramming and communication linking. The concept is adapted from plug and play from the office world. In the literature there are a few approaches showing that plug and play cannot easily be converted to plug and produce. In a manufacturing system one needs to deal with complex machines, which are not standardised and often have mechanical interference or share the same workspace. That increases the complexity of adding. The range of capabilities are too different, a component can be simple like a sensor or complex as a robot. On top of this the plug and produce concept must still provide real time behaviour for the system and fulfil bus system timing and bandwidth requirement. Also an important condition for the industrial-level is robustness and the health and safety regulations which must match the company's policy [10].

In the literature review different approaches of how and why researchers have investigated plug and produce can be found. One motivation is to reduce the integration effort when a production layout has changed [11]. A method is developed to standardise software components and machines. Important data about machines are stored in a database to automatically plan if a new machine layout is required. While in [12] the integration time is reduced with a method to automatically calibrate robots and allocate workspace based on the task when a new tool is installed. Another approach involves changing the machines of a production line e.g. parts of robots. In [13] a robotic system is proposed to build and program from parts out of a CAD catalogue which includes the functionalities and limits. The drawback of this approach is the dependency on specialised hardware to realise the integration and information about the parts which can be difficult to get from different vendors.

Another approach is to minimise the integration effort by introducing a control architecture that contains commands for machine parts, like manipulator etc. [14]. If a part is plugged in commands are shown to the operator to make the programming less complex for less trained persons. In [15] a concept to store configurations and functions in EDDL (electronic device description language) is presented. The system needs to be configured only once and it can be reused from the central data management. However, these approaches lack of a dynamic adaptation to a changing environment and complex demonstration scenarios.

In [1] the procedure of plug and produce is presented and used in combination with autonomous and cooperative systems, called holons or agents. The system is located in the computer network and checks the condition of the machines frequently. This allows a high level management of the assembly machines and an experiment shows that a replacement of a machine can be realised in a shorter time. In [16] the experiments and products to assemble are extended. The automatic calibration of the robotic system is included and controlled by the agent system.

The interested reader is directed to [17, 18] for a general introduction to multi-agent systems (MAS).

An extension to the management of plug and produce by the agents is described in [19], where an agent-based control system with diagnostic possibilities is presented. Based on Hidden Markov Models the authors introduce several layers for diagnostic of the system. Whereas a lower layer monitors the control hardware, a higher level does diagnostics and classifies types of failure. This system is able to collect and detect failures more often. Based on this, further work could include automatic failure recover and decision-making support for the user.

As we have seen, there is a need for customised products and a quickly response to changes on the market. The literature review identified plug and produce as a key element [1] to integrate new devices in order to configure machines with minimum reconfiguration effort. In combination with multi-agent technology properties like autonomy, openness and communication features, this will extend the plug and produce concept to deal with workflow changes and automatic task assignment [2].

There is no architecture available to enable plug and produce capabilities with standard technology. To provide a basis for the complex structures and communication of the reconfigurable production system, we propose a multi layered multi-agent architecture for plug and produce. This architecture will extend an existing setup to provide self-reconfiguring capabilities. We will show the benefits of this contribution by implementing the architecture on a real industrial demonstrator.

2   The Multi-Agent Architecture

This section introduces the production environment in which agents will be deployed. Agents are defined and specified as part of the architectural design. The plug and produce system is designed as a multilayered multi-agent system, where each layer is represented as a graph of agents [20] [21]. The benefits of a layered multi-agent architecture are visible, for example, in resource bounded agents with a clear methodology and a modular structure [22].

In Figure 1 a generic production system which is able to produce a final product is shown. The architecture is an extension to the traditional system. The parts in grey indicate necessary additions to host the agent system in order to enable plug and produce capabilities. Attached to each controller is a machine. The order of machines is given by the product. In case of changing the production system layout the PLC and the machine will be moved together.

The architecture in Figure 1 is based on the three layers of a traditional production system:

1. **HMI Layer**: this layer interacts with the end user by receiving input from the operator and delivering information to the agents. The goal of this interaction is to control and supervise the production system.

2. **Control Layer**: this layer controls the production system by monitoring the production process through sensors and by commanding the execution of actions through actuators.
3. **Production System Layer**: this layer contains the production system where machines and the material handling devices operate to produce the desired product

In the following, the agent types are described. The ***Component Agent*** (CA) can be deployed either on a PLC or on a small processing unit which will then be connected to the PLC (both shown in Figure 1). The latter enables the use of legacy systems. The ***Plug and Produce Management Agent*** (P&PA) manage the plug and produce activities by monitoring the component agents and interfacing with the HMI. These agents are dormant during the operation of a production system, yet aware of component agents plugged in. However, they become active whenever a plug and produce activity is initiated. Once a module is plugged in and detected by the P&PA, the configuration for this device will be uploaded from the database (DB) to the relevant controller.

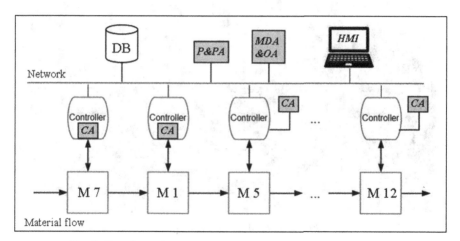

Fig. 1. Generic production system without and with the agent system

Afterwards the ***Monitoring, Data Analysis and Optimisation Agent*** (MDA&OA) become active to police the system. Additionally, they are used to monitor performance indicators and perform data analysis for process optimisation. Monitoring information can be displayed on the HMI to be used directly by the operator or written to a database to perform an analysis separately.

The management of the machines is done by the agents through the existing communication link. The process control activities are left to the control layer to avoid collisions with machines. If the agent system fails plug and produce capabilities are disabled but the production system continues to operate. This level of robustness is necessary for industrial implementation because of the reduced risk of stopping the production which normally is connected to high costs.

3   Implementation

In order to demonstrate our idea we use a real industrial production cell from Feintool Automation, shown in Figure 2, which is suitable to address all our research questions. This type of cell is used, for example, in some European factories to assemble photovoltaic panels. The production cell contains two robots, tool changing equipment (not displayed in the figure) and a testing station with an adaptable hexapod. At the bottom there are eight boxes with industrial used PLCs inside, one for each workspace. A shuttle system is serving all workstations.

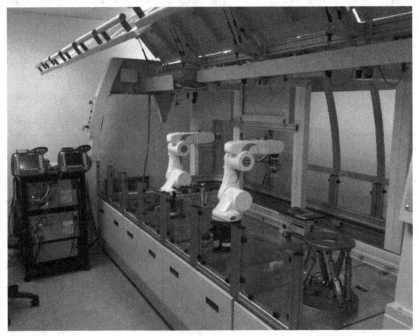

Fig. 2. Testbed for plug and produce

Figure 3 is the top view of Figure 2 after setting up.

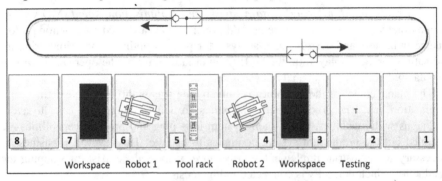

Fig. 3. Demonstrator setup for plug and produce

We are currently working on the following use case to demonstrate plug and produce and self-reconfiguration.

For this scenario the robots need to have both access to the same tool changing rack to be able to assemble the full product.

Firstly, robot 1 is deactivated. Robot 2 assembles and tests the product and is monitored by the MDA&O agent type. Performance data is displayed on the HMI. The operator decides to plug in robot 1 to produce in time. As soon as robot 1 is plugged in, the P&PA detects a new machine and loads the corresponding configuration from the DB. After integration the responsible agents become dormant again. The MDA&O agents are notified about the changes and execute the optimisation algorithm. When a better configuration is found the operator is asked to confirm the change on the HMI. On this demonstrator the configuration of the shuttle system will be modified to serve robot 1. Also, depending on the robot capabilities (e.g. speed, energy consumption, precision) both robots will share assembly tasks in order to achieve the best result.

In Figure 4 is an example of assembly of three items of a product. The assembly consists of three steps (1,2,3) which can be performed by either robot. The parts are carried on a pallet which can be moved by the robots. In the figure one robot is labelled as A, the other is marked as B. A2 stands for the second assembly step by the first robot. The parts and the final products are transported by a shuttle system indicated by the letter 'S'. A testing procedure is carried out at the end of each assembly and labelled by the letter 'T'. It is assumed, for the sake of simplicity of the drawing, that all tasks take the same amount of time.

As you can see in the table at the top of Figure 4 the shuttle needs to go two steps towards robot A which is in this case located after robot B. After the three assembly steps are performed by robot A, the product is tested and transported away by the transport system. Then the next shuttle with product parts comes to start again the process for the next product.

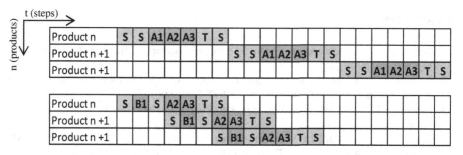

Fig. 4. Task sequence diagram to produce three of the same products

The table at the bottom of Figure 4 visualises the task sequence for two robots. Now the shuttle goes first to robot B in order to perform the first assembly step. The way and number of steps from the shuttle does not change so the shuttle goes to robot A to finish the product. Once the robot A takes the pallet from the shuttle, the pallet is empty and ready to serve robot B in order to work in parallel. With three products and the given setup this would bring a saving of 62 % of time compared to the former case.

4   Conclusions

One way to reduce the effort of reconfiguration tasks is to implement self* techniques for machines [23], e.g. to be self-aware of the machine status or self-organise the order of jobs. One technology identified by the literature is the multi-agent system where multiple agents communicate to reach an agreement that satisfies their individual goals [17, 24]. The reasons why multi-agent technology is not fully accepted in manufacturing are listed in [2]. Our work will focus on the main barriers for an industrial take-up of multi-agent systems to implement plug and produce. In various European projects (e.g. EUPASS, IDEAS) the hardware was modified in order to host the agent system. We use standard technology to host the agent system. This allows the operator to get support by the vendor and the availability is secured in opposition to modified controller. Another barrier we focus on is the integration of legacy systems, with the objective of avoiding high investments to replace hardware. The architecture will be deployed on existing hardware or little processing units which host the agents and work as a middleman.

Another barrier for industry is the potential unpredictability of actions performed by self-interested agents to reach their goal. Our contribution is an extension to the traditional production control, in the sense that if the agent system fails the plug and produce capabilities will stop but the production system will continue to operate. Such a fault-tolerant system will be supported by our architecture by separating the production line into layers. This behaviour will reduce the risk for industries to have the whole production stopped because of a failure.

We will extend the architecture and validate it on the testbed with different use case scenarios. Furthermore, we will develop a generic methodology for plug and produce to enable the self-reconfiguration of the system.

References

1. Arai, T., et al.: Agile assembly system by 'Plug and Produce'. CIRP Annals - Manufacturing Technology 49(1), 1–4 (2000)
2. Monostori, L., Váncza, J., Kumara, S.R.T.: Agent-based systems for manufacturing. CIRP Annals - Manufacturing Technology 55(2), 697–720 (2006)
3. Reinhart, G., et al.: Automatic configuration (Plug & Produce) of Industrial Ethernet networks, Sao Paulo (2010)
4. Sauer, O., Jasperneite, J.: Wandlungsfähige Informationstechnik in der Fabrik. ZWF: Zeitschrift für Wirtschaftlichen Fabrikbetrieb 105(9), 819 (2010)
5. Furmans, K., Schönung, F., Gue, K.R.: Plug-and-Work material handling systems. In: 2010 International Material Handling Research Colloquium, Milwaukee, USA (2010)
6. Sauer, O., Ebel, M.: Plug-and-work of production plants and superordinate software. In: Plug-and-work von Produktionsanlagen und Übergeordneter Software, Bremen (2007)
7. Lepuschitz, W., et al.: Toward Self-Reconfiguration of Manufacturing Systems Using Automation Agents. In: IEEE Transactions on Systems, Man and Cybernetics Part C: Applications and Reviews (2010)
8. Marik, V., McFarlane, D.: Industrial adoption of agent-based technologies. IEEE Intelligent Systems 20(1), 27–35 (2005)

9. Mařík, V., et al.: Rockwell automation agents for manufacturing. In: Proceedings of the Fourth International Joint Conference on Autonomous Agents and Multiagent Systems. ACM (2005)
10. Zimmermann, U.E., et al.: Communication, configuration, application - The three layer concept for Plug-and-Produce, Funchal, Madeira (2008)
11. Sauer, O., Ebel, M.: Plug-and-work von Produktionsanlagen und übergeordneter Software. Aktuelle Trends in der Softwareforschung, Tagungsband zum do it. software-forschungstag, pp. 24–33 (2007)
12. Maeda, Y., et al.: "Plug & Produce" functions for an easily reconfigurable robotic assembly cell. Assembly Automation 27(3), 253–260 (2007)
13. Wögerer, C., Bauer, H., Rooker, M., Ebenhofer, G., Rovetta, A., Robertson, N., Pichler, A.: LOCOBOT - low cost toolkit for building robot co-workers in assembly lines. In: Su, C.-Y., Rakheja, S., Liu, H. (eds.) ICIRA 2012, Part II. LNCS, vol. 7507, pp. 449–459. Springer, Heidelberg (2012)
14. Naumann, M., et al.: Robot cell integration by means of application-P'n'P. VDI BERICHTE 1956, p. 93 (2006)
15. Reinhart, G., Hüttner, S., Krug, S.: Automatic configuration of robot systems-upward and downward integration. In: Intelligent Robotics and Applications, pp. 102–111. Springer (2011)
16. Arai, T., et al.: Holonic assembly system with Plug and Produce. Computers in Industry 46(3), 289–299 (2001)
17. Wooldridge, M.: An introduction to multiagent systems. Wiley.com (2009)
18. Bussmann, S., Jennings, N.R., Wooldridge, M.: Multiagent systems for manufacturing control: a design methodology. Springer (2004)
19. Lepuschitz, W., et al.: A Multi-Layer Approach for Failure Detection in a Manufacturing System Based on Automation Agents. In: 2012 Ninth International Conference on Information Technology: New Generations (ITNG). IEEE (2012)
20. Bandini, S., Manzoni, S., Simone, C.: Dealing with space in multi–agent systems: a model for situated MAS. In: Proceedings of the First International Joint Conference on Autonomous Agents and Multiagent Systems: Part 3. ACM (2002)
21. Weyns, D.: Architecture-based design of multi-agent systems. Springer (2010)
22. Fischer, K., Müller, J.P., Pischel, M.: Unifying control in a layered agent architecture (2011)
23. Babaoglu, O., et al.: The self-star vision. In: Self-star properties in complex information systems, pp. 1–20. Springer (2005)
24. Markus, A., Kis Vancza, T., Monostori, L.: A market approach to holonic manufacturing. CIRP Annals-Manufacturing Technology 45(1), 433–436 (1996)

Process Module Construction Kit for Modular Micro Assembly Systems

Raphael Adamietz[*], Tobias Iseringhausen, and Alexander Verl

Fraunhofer-Institute für Produktionstechnik und Automatisierung IPA,
Nobelstr. 12, 70569 Stuttgart, Germany
adamietz@ipa.fraunhofer.de
www.ipa.fraunhofer.de

1   Introduction

The production environment in Europe is characterized by a number of different challenges. Production volumes are hardly predictable due to the turbulent market [1]. Product life cycles become increasingly shorter [2]. Customers demand more individual and customized products, which leads to increasingly diversified product pallets [3]. The occurrence of changes on product and process, even for running productions, increases continuously [4]. At the same time the field of tension between assurance of quality, reduction of lead times and reduction of cost remains present [5].

The trend of miniaturization with the concurrent trend of integration of functions is a central topic in the product development in different fields of application [6], like optics, electronics, medical technology, bio technology, information technology and aeronautical technology. With ongoing miniaturization many companies face challenges of precision and ultra-precision assembly technology. Therewith increases the cost and complexity of the required process equipment and the demands to skilled personnel [7].

The development of the markets in micro system technology shows that many new and innovative developments in the field of hybrid microsystems cannot achieve a satisfactory success [8]. This can be attributed to the high complexity of microtechnical products and processes, a lack of interdisciplinary knowledge in process development, limited flexibility of the applied manufacturing and assembly systems and the high investment risk due to uncertain forecasts of growth [9]. From the economic point of view, often only mass production is reasonable [9].

Modular micro assembly systems are often regarded as suitable link between the requirements of the market environment and the current state of development in micro production technology [10-12]. Approaches such as the Agile Assembly Architecture [11], Evolvable Assembly Systems [10] or the Reconfigurable Micro Assembly System [12] are parts of ongoing research and have been partly transferred to commercial products.

Modular production systems are characterized by standardized components and defined interfaces [13]. Single modules can be exchanged rapidly and re-installed in

---

[*] Corresponding author.

S. Ratchev (Ed.): IPAS 2014, IFIP AICT 435, pp. 126–132, 2014.

other systems. This provides many advantages over the whole product life cycle. In the field of micro production these effects are increased, due to the increased investment cost and higher complexity of the production equipment.

However, the market share of modular machines does not match the predictions [14]. This can be mainly attributed to the fact that the initial investment into a modular machine is significantly higher than into a special purpose machine. Repetition effects by re-use of equipment in form of process modules are necessary to make modular machines more attractive.

The authors of this paper suggest an open interface which allows the integration of process equipment of different providers into machinery of different providers. This would significantly improve the possibility to continuously adapt assembly machines. Particularly companies with broad product ranges could potentially benefit strongly.

Based on this open interface a common platform for the exchange of process modules is suggested [15]. By a certification step, the compatibility of the offered process modules and platforms is ensured. Providers and users of modular production equipment can then offer and exchange their equipment. This way the application of one process module is not bound to one platform, but can be transferred across different platforms of different providers. This would have a strong effect on the availability of process equipment.

To support the development and set-up of process modules, a construction kit is proposed. Within this paper, the basic concept of a construction kit for process modules is described. Furthermore examples of configurations of process modules based on the construction kit are presented.

2   Qualitative and Quantitative Analysis

The approach of modular micro assembly machines has been compared to state-of-the-art micro assembly machinery in previous work. For qualitative analysis a SWOT[1] analysis was presented and for quantitative analysis a net present value analysis was performed. [15]

The quantitative analysis showed potential for risk management by the application of a modular system. If for example the production volumes deviate from the predictions, the modular approach allows adapting to these changes.

The qualitative analysis revealed many opportunities such as simultaneous engineering, standardized modules which can be used already during process development and reuse of production equipment. However, the analysis also revealed several threats, like decrease of process stability and product quality which might be caused by the changes of the assembly system. The largest threat, however, was that process modules might not be available as required. As an approach to improve this situation, the concept of a process module construction kit is presented here.

---

[1] SWOT – Strength, Weaknesses, Opportunities, Threats.

3   Concept of a Process Module Construction Kit

3.1   Complex Interface

Also proposed within previous work, was the 'Complex Interface' [15]. It consists of three concepts: The 'mechanical interface', the 'supply interface' and the 'control interface'.

The 'mechanical interface' comprises the definition of a generic, blank module. Spaces for process, feeding and control equipment as well as working spaces are defined. Within the 'supply interface' a supply connector is defined which includes supply of compressed air and electricity as well as connection to Ethernet and field bus. The 'control interface' defines the interface and function blocks for control integration. The overall complex interface definition is a prerequisite for cross-platform integration. It allows machine constructors to provide a compliant interface to a common process module specification. Furthermore, it allows third parties to develop and to build process modules for different machines providing the complex interface.

3.2   Process Module Construction Kit

The effort required to develop and to set up process modules could be potentially reduced by a construction kit. A concept for such a construction kit has been designed and is presented here. The design process is depicted in Fig. 1 and Fig. 2.

It is assumed the requirements and boundary conditions have been determined and the process equipment has been chosen as the pre-condition for application of the construction kit (1). Process equipment is here referred to as e.g. a dispenser or a gripper.

Within the next step it is checked if the chosen process equipment can be integrated to a process module according to the specifications (2). This comprises the mechanical specifications and interfaces, the supply of the process module and an integration of the control system. Many process module systems work with different pitches, making the choice of an applicable process module necessary (3).

If the result of the check against the specification is positive, the process module can be designed. It is the aim to provide as many pre-defined elements as possible with known properties to ease the design process. These 'reference solutions' (4) comprise components required to build up the process module, such as axis systems with different kinematics (xy-tables, xyz-systems, 6-axis-kinematics, ...), pre-defined portal elements, sensors and calibration equipment (TCP-measuring units, vision systems, ...), feeding equipment (waffle pack feeder, vibrational feeder) and supportive equipment (cleaning devices, trays). With this equipment a process module, based on reference solutions can be designed (5).

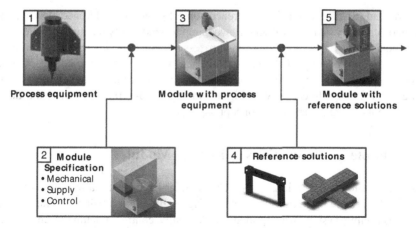

Fig. 1. Process module construction kit based on reference solution

To support the design and the set-up process a mechatronic tool interface is provided (6). It allows instant integration of process equipment, like e.g. a gripper, a dispenser or a camera system. While the complex interface provides interfacing between process module and machine, the mechatronic tool interface provides a standard interface between equipment on the process module itself. The interface is also comprehensive and provides a mechanical, supply and control interface. However, the specifications necessary for a tool interface differ from those of a module interface. The applied interface could be based for example on an ISO 29262 tool exchange interface [16]. An adaption to feed through field bus, Ethernet and media is additionally required.

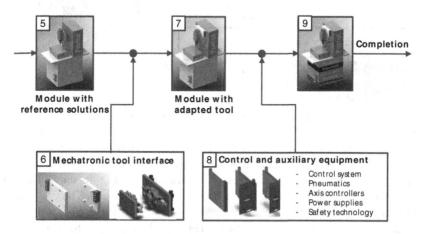

Fig. 2. Process modules based on reference solution

Once the process module with adapted tool is designed (7), control and auxiliary equipment are required. Reference solutions for control and auxiliary equipment are also provided (8). This comprises reference control programs, reference electrical and pneumatic circuitry, axis controllers, power supplies and safety technology.

The next step is the completion of the process module (9). This includes the design, manufacturing and set-up of the parts which are not covered in the construction kit. Furthermore testing is necessary before commissioning. Although the elements of the kit are depicted in a linear way in Fig 1 and Fig 2, re-iterations during the design phase are possible and part of the concept.

4   Sample Configurations of Process Modules

Different sample configurations of process modules have been realized as CAD models. One of these configurations is shown in Fig. 3. Based on the third module variant, a xyz-positioning system was applied. Additionally control, pneumatic and electric circuitry solutions are provided from the construction kit.

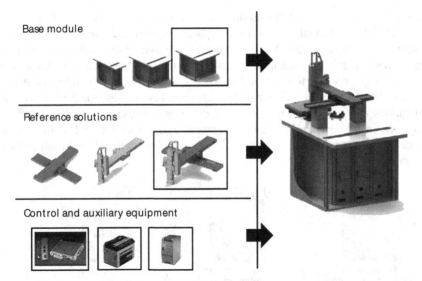

Fig. 3. Sample configuration of a process module

Regarding the adaption of tools, three different configurations are depicted in Fig 4. The adaption of a vision system together with a precision vacuum gripper is shown in Fig 4a. This could be applied for a pick and place process. Fig 4b shows the same gripper combined with an extendible time-pressure dispensing unit. The combination could be used for a precision assembly operation using an adhesive. Fig 4c shows the dispenser in conjunction with the camera. This could be applied for a dispensing process supported by machine vision.

Although the number of elements in the construction kit is still very limited, many different reasonable combinations turned out to be possible. Still in the state of a model, first experience promises significant reduction of effort for the design of a process module.

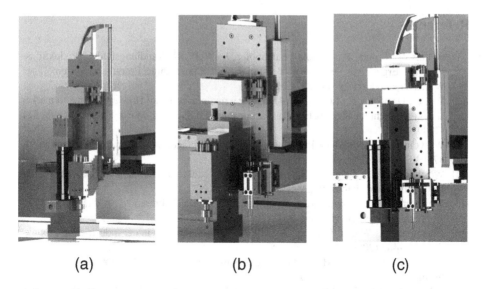

(a)   (b)   (c)

Fig. 4. Sample configurations of process modules based on the process module construction kit

5   Conclusions and Outlook

A concept for a construction kit for process modules is presented in this paper. The principle and the basic elements were described. Aim is to reduce the effort for design and set-up of process modules. The approach is to provide as many pre-defined elements as possible and to provide standard interfaces. However, it turned out that room for individual adaptions needs to be foreseen, as different applications have different requirements and that a one-fits-all solution is hardly achievable. Therefore the construction kit provides a pool of elements, which the developer can choose from and still provides room for individual design towards the specific requirements of the application.

To test the construction kit, a number of reference solutions were modeled. Based on these elements a number of different process modules were realized in CAD. The effort during modeling was significantly reduced by access to the pre-defined elements. However, the effect was not compared to commercially available solutions yet.

A prototype of a process module based on the construction kit will now be realized to evaluate the concept. The results will be used to re-iterate and optimize the concept and the included elements. An extension towards adaption of already existing solutions to new requirements will be included. The framework will furthermore be extended by supportive software tools for the construction kit and the specification micro assembly tasks.

Acknowledgements.   This   paper   has   emerged   from   the   project «PRONTO/VOLPROD», which has been sponsored by the German Federal Ministry of Education and Research (BMBF) in the funding program "SpitzenCluster Micro-Tec SüdWest" and is administered by the project sponsor VDIVDE-IT in Berlin.

References

[1] Nyhuis, P., Fronia, P., Pachow-Frauenhofer, J., Wulf, S.: Wandlungsfähige Produktions- systeme: Ergebnisse der BMBF-Vorstudie "Wandlungsfähige Produktionssysteme". wt Werkstattechnik online 99, 205–210 (2009)

[2] Wu, L., De Matta, R., Lowe, T.J.: Updating a modular product: How to set time to market and component quality. IEEE Transactions on Engineering Management 56, 298–311 (2009)

[3] Westkämper, E.: Modulare Produkte - Modulare Montage. wt Werkstattechnik Online 91, 479–482 (2001)

[4] Westkämper, E., Zahn, E.: Wandlungsfähige Produktionsunternehmen: Das Stuttgarter Unternehmensmodell. Springer (2008)

[5] Westkämper, E.: "Produktion im turbulenten Umfeld. Presented at the ATZ/MTZ- Konferenz AutomobilMontage 2007 (2007)

[6] Hesselbach, J., Wrege, J., Raatz, A.: Mikromontage. Montage in der industriellen Produk- tion, 463-482 (2006)

[7] Popa, D.O., Stephanou, H.E.: Micro and mesoscale robotic assembly. Journal of manufac- turing processes 6, 52–71 (2004)

[8] Schilp, J.: Adaptive Montagesysteme für hybride Mikrosysteme unter Einsatz von Telepräsenz, vol. 244. Herbert Utz Verlag (2012)

[9] Hesselbach, J., Raatz, A., Wrege, J., Herrmann, H., Weule, H., Buchholz, C., Tritschler, H., Knoll, M., Elsner, J., Klocke, F.: Untersuchung zum internationalen Stand der Mikro- produktionstechnik

[10] Onori, M.: Final Report - EUPASS (Evolvable Ultra-Precision Assembly Systems), http://cordis.europa.eu/documents/documentlibrary/127976311E N6.pdf2011

[11] Muir, P.F., Rizzi, A.A., Gowdy, J.W.: Minifactory: A precision assembly system adapta- ble to the product life cycle. In: Intelligent Systems & Advanced Manufacturing, pp. 74–80 (1997)

[12] Gaugel, T., Dobler, H., Rohrmoser, B., Klenk, J., Neugebauer, J., Schäfer, W.: Advanced modular production concept for miniaturized products. In: Proc. of 2nd International Workshop on Microfactories, Fribourg, Switzerland, pp. 35–38 (2000)

[13] Nyhuis, P., Heinen, T., Reinhart, G., Rimpau, C., Abele, E., Wörn, A.: Wandlungsfähige Produktionssysteme: Theoretischer Hintergrund zur Wandlungsfähigkeit von Produk- tionssystemen. wt Werkstattstechnik online 98, 85-91 (2008)

[14] Kluge, S.: Methodik zur fähigkeitsbasierten Planung modularer Montage systeme (2011)

[15] Adamietz, R., Iseringhausen, T., Gerstenberg, S., Verl, A.: A First Step towards Cross- Platform Integration in Modular Micro-assembly Systems–Concept for a Process Module Construction Kit. In: Enabling Manufacturing Competitiveness and Economic Sustaina- bility, pp. 35–40. Springer (2014)

[16] Deutsches Institut für Normung, DIN ISO 29262 - Fertigungsmittel für Mikrosysteme - Schnittstelle zwischen Endeffektor und Handhabungsgerät. Beuth Verlag

Modular Workpiece Carrier System for Micro Production

Tobias Iseringhausen, Raphael Adamietz, Dirk Schlenker, and Alexander Verl

Fraunhofer Institute for Manufacturing Engineering and Automation IPA, Stuttgart, DE

Abstract. In micro production high lot sizes, which are required for cost-effectiveness, are often not achieved, e.g. due to a high number of product variants or limited demand. Therefore new approaches for production like changeable manufacturing systems are demanded. In this paper a concept for a modular workpiece carrier system including a platform for sensors and actors is presented. The modular design allows the easy adaption of the manufacturing system to new products by the workpiece carrier system in terms of hardware and even the integration of additional process functionality.

Keywords: changeability, micro production, precision assembly, workpiece carrier.

1   Introduction

While product life cycles get shorter, production volumes vary and the number of product variants increases [1]. In some branches, especially the manufacturing of hybrid micro systems like sensors and other electro mechanical devices, nowadays often only low to medium lot sizes are reached, due to limited demand. As the conventional automated production with special purpose machines requires high lot sizes to be profitable, new approaches for efficient production are required [2]. Another aspect beside the lot sizes is the limited capability of manufacturing systems regarding additional functionality. For example, the integration of additional actors and sensors requires a large effort in terms of time and money, especially for the control adaption of the often closed systems, and causes related downtimes.

To cope with the changing requirements that the market imposes, increasingly changeable systems are required [3]. Further functionality is often provided by additional processes in the form of process modules. However, an increase of changeability can also be achieved by other system components. Generally, by improving system properties (changeability enablers) towards modularity, compatibility, mobility, universality and scalability, the changeability of the overall system is improved [3]. This applies to all levels of the factory, including the components level. One component, which is strongly linked to both the product and the production system, is the workpiece carrier. An increase of changeability of the workpiece carrier would improve the overall system changeability and offer a further possibility to adapt to product and process requirements. Furthermore, process functionality could be added to the

S. Ratchev (Ed.): IPAS 2014, IFIP AICT 435, pp. 133–138, 2014.

workpiece carrier, which could enable further possibilities of system evolution on many levels.

2   Related Work

To increase changeability and enhance process functionality on component level by workpiece carriers, corresponding systems have been developed. For example, to manufacture 3D-MID devices in conventional PCB manufacturing lines a multi axis workpiece carrier has been developed [4]. For applications in milling machines an automated, reconfigurable fixture has been developed, which allows automated fixture planning and automated fixture configuration. It is equipped with sensors for control. Exchangeable jaws enable the reconfiguration [5]. To suppress self-induced vibrations during cutting on machine tools and thus increase surface quality and reduce shape deviations, active workpiece carriers with mostly piezoelectric actuators are a current research topic [6, 7]. Due to the purpose of workpiece carriers, it is a wide-ranging topic with systems often adapted to very special purposes.

3   Modular Workpiece Carrier System

In this paper a concept of a workpiece carrier system designed for the application in micro production processes is presented. There is a focus on micro assembly, and therefrom requirements are derived. Micro assembly systems and micro assembly tasks are distinguished by some typical boundary conditions compared to conventional macro assembly. To determine these boundary conditions, a survey targeting microsystem manufacturers in Germany has been performed [8].

The most obvious difference between micro and macro is the size of microtechnical products. 87% of microtechnical devices fit into a volume of 100 x 100 x 100 mm³, while 30% are even smaller than 10 x 10 x 10 mm³ [8]. Small structures and functional elements demand positioning accuracies between 5 - 25 microns or even higher during assembly [8]. This imposes high requirements on the overall assembly system, like stiffness and precision, and the positioning technique itself. Furthermore, increased demands on cleanliness and defined, constant temperatures have to be regarded. Another characteristic is the high number of product variants. 75% of microtechnical products have up to ten or more variants [8].

The tracking of quality relevant parameters gains increasing importance. This applies particularly to many medical products, but the tracking requirements also exist in other industries (automotive, aerospace, etc.). Furthermore for high quality products, the data obtained by extensive tracking and tracing can be used for production optimization.

To comply with these requirements and increase the changeability and process functionality as described above, the concept of the workpiece carrier system is modularly designed to allow the adaption of the workpiece carrier by the easy exchange of functional elements like sensors and actors, so called smartFeatures. It consists of a modular electronic platform which is matched with a modular mechanical design.

This allows a broad range of applications in micro production, nevertheless by the scalability of the system and operating distances even large scale systems are possible.

3.1   Modular Electronic Platform

The main items of the modular electronic platform are sketched in figure 1 below. The central component is a control unit, which is integrated into the mobile workpiece carrier. It allows data acquisition, processing and output of the connected sensors and actors. The communication within the workpiece carrier is done wireless if beneficial or tethered. The communication to the superior system, e. g. the machine control, is done wireless via a machine coupler, which is equipped with machine adapted control interfaces.

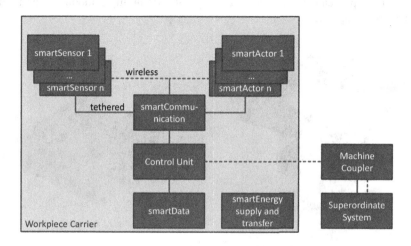

Fig. 1. Modular electronic platform

For energy supply a battery system and an energy transfer system with capability of wireless power is provided. Beside the control unit and machine coupler the further components are designed as modular. Thus it is possible to adapt the system to the requirements and conditions of the particular use case. For instance energy-saving smartFeatures combined with a high-capacity energy supply can be used for long-term applications with reduced processing power requirements. The smartSensors are capable of data acquisition and preprocessing to, for instance, reduce data volume within the electronic platform or increase accuracy by preprocessing close to the sensor to avoid signal interferences. Different types are drafted: one type with capability of high frequency applications, one low power type and one type designed for high accuracy.

3.2   Modular Mechanical Design

The associated mechanical design of the workpiece carrier system is consistently modularly designed, see figure 2 below. The central element of the mechanical design is a workpiece carrier with frame architecture, see top element in the figure above. To facilitate the dissemination of the system, the outer dimensions and the outer mechanical interfaces refer to the standard DIN 32561. This standard describes dimensions and tolerances of a tray in the field of production equipment for microsystems. The mandatory elements of the electronic platform, control unit, communication as well as energy supply and transfer, are placed into the frame. Thus an encapsulation of the electronic components in rough environments like ultra-precision machining is ensured. The aperture within that frame architecture can now be individually designed with modules, considering the mechanical and electronic interfaces to the workpiece carrier and is matched to the size of 100 x 100 mm², which most microtechnical devices fit into, pursuant to [8]. For the integration of multiple modules distribution modules can be applied.

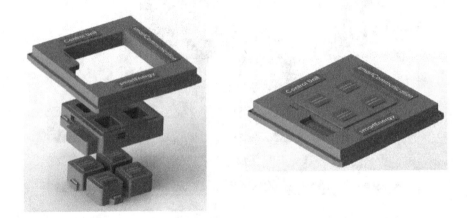

Fig. 2. Mechanical design of the workpiece carrier system

By flexible control interfaces of the machine coupler and even simple ones, like digital inputs and outputs, the system can be integrated into almost every machine with minor effort. The setup and programming of the workpiece carrier can take place outside the machines in specially designed set-up stations if required. Due to the compact design with low height and different sizes the mechanical integration is feasible in many, even limited, working spaces.

To give an illustrative draft of a workpiece carrier system, one example adapted for precision machining is sketched in the figures below.

Here modules with motor-driven clamping are inserted into the workpiece carrier. Additionally there is a module with inertial measurement unit and temperature sensor. It allows the measurement of the environmental temperature close to the workpiece and the detection of shocks during the manufacturing of sensitive parts. In the event of varying workpieces, for example, the clamping module can be easily changed without modification of the machine tool.

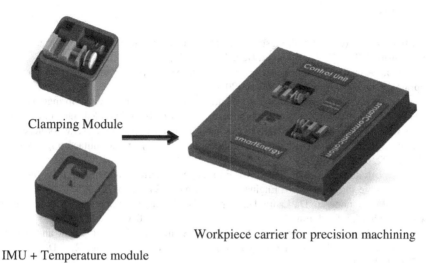

Clamping Module

Workpiece carrier for precision machining

IMU + Temperature module

Fig. 3. Components of workpiece carrier for precision machining

4   Discussion

Compared to other work in the field of workpiece carriers in this paper a concept for a flexible and modular system, adapted to the requirements in micro assembly and re-lated micro production processes requiring high precision is presented.

Beyond conventional passive workpiece carriers the modular electronic platform enables the integration of process and monitoring functionality. Its own energy supply allows, combined with the wireless data transfer capability, the continuous track and trace in production of quality and safety related devices. Due to the energy supply also the acquisition of measurement data or the control of actors is possible, even outside machines during transport or storage.

The modular design enables the precise adaption to the product and processes without dissipation of resources by overdesigned or unutilized features of the work-piece carrier. The standardization of the modules and their interfaces simplifies the adaption and makes reusability of modules possible.

The next tasks will be the implementation of the single elements of the modular electronic platform and the modular mechanical design as well as the merging to workpiece carrier systems. These systems will be technically and economically vali-dated in several industrial use cases in micro assembly, ultra-precision machining and precision measurement.

Acknowledgements. This paper has emerged from the project »smartWT«, which has been sponsored by the German Federal Ministry of Education and Research (BMBF) in the funding program "Spitzencluster MicroTEC Südwest" and is administered by the project sponsor VDI/VDE-IT in Berlin.

References

[1] Westkämper, E. (ed.): Wandlungsfähige Produktionsunternehmen: Das Stuttgarter Unternehmensmodell, pp. 7–23. Springer, Heidelberg (2009)

[2] Schilp, J.: Adaptive Montagesysteme für hybride Mikrosysteme unter Einsatz von Telepräsenz, vol. 244. Herbert Utz Verlag (2012)

[3] ElMaraghy, H.A.: Changeable and reconfigurable manufacturing systems. Springer (2009)

[4] Pfeffer, M., Goth, C., Craiovan, D., Franke, J.: 3D-Assembly of Molded Interconnect Devices with standard SMD pick & place machines using an active multi axis workpiece carrier. In: 2011 IEEE International Symposium on Assembly and Manufacturing (ISAM), May 25-27, pp. 1–6 (2011)

[5] Shea, K., Ertelt, C., Gmeiner, T., Ameri, F.: Design-to-fabrication automation for the cognitive machine shop. Advanced Engineering Informatics 24(3), 251–268 (2010)

[6] Brecher, C., Manoharan, D., Ladra, U., Köpken, H.-G.: Chatter suppression with an active workpiece holder. Prod. Eng. Res. Devel. 4(2-3), 239–245 (2010)

[7] Abele, E., Hanselka, H., Haase, F., Schlote, D., Schiffler, A.: Development and design of an active work piece holder driven by piezo actuators. Prod. Eng. Res. Devel. 2(4), 437–442 (2008)

[8] Adamietz, R., Iseringhausen, T., Schlenker, D.: VolProd - Survey among microsystem manufacturers in Germany. In: Fraunhofer IPA (2012)

A Generic Systems Engineering Method for Concurrent Development of Products and Manufacturing Equipment

Erik Puik[1], Paul Gielen[2], Daniel Telgen[1], Leo van Moergestel[1], and Darek Ceglarek[3]

[1] HU University of Applied Sciences Utrecht, Oudenoord 700, 3513EX Utrecht,
The Netherlands
[2] MA3Solutions BV, Science Park 5080 Eindhoven, 5692 EA, The Netherlands
[3] International Digital Laboratory, WMG, University of Warwick, Coventry, CV4 7AL, UK
erik.puik@hu.nl

Abstract. Manufacturing is getting more competitive with time due to continuously increasing global competition. Late market introduction decreases the economic lifecycle of products and reduces return on investments. Reconfigurable Manufacturing Systems (RMS) reduce the time to market because the process of equipment configuration is less time consuming than engineering it from scratch. This paper presents a scientific framework, to be applied as an engineering design tool, that is capable of improving the relation between product design and the reconfiguration process of RMS. Not only does it support a cross-domain adjustment and information exchange between product developers and manufacturing engineers, it also adds risk analysis for conscious risk taking in a cyclic development process. The method was applied on an industrial case; concurrent design and manufacturing of an environmentally friendly circuit board for wireless sensors. The method may be considered successful. It will lead to better system architecture of product and production systems at a more competitive cost. Feedback on the development process comes available in the early development stage when the product design is not rooted yet and two-way optimisations are still possible.

Keywords: Reconfigurable Manufacturing System, RMS, Risk modelling, FMEA, Qualitative Analysis, Concurrent Engineering, Agile Manufacturing, Production, Equiplets, Micro, Hybrid, Microsystem.

1 Introduction

Increased global competition in manufacturing technology puts pressure on lead times for product design and production engineering. Quickly eroding markets, like markets for high-tech systems and micro-technology, especially require tight scheduling of system development; 'being first' leads to an extended economic lifecycle, better market penetration and higher added value for products, together leading to progressively higher return on investments [1]. By the application of effective methods for systems engineering (engineering design), product design and production development can be executed in parallel instead of sequentially. Modular equipment is currently under developed to not only meet the manufacturing demand of single products but to address product groups [2, 3]. Instead of developing dedicated

S. Ratchev (Ed.): IPAS 2014, IFIP AICT 435, pp. 139–146, 2014.
© IFIP International Federation for Information Processing 2014

manufacturing systems for specific needs, Reconfigurable Manufacturing Systems (RMS) reuse modular parts of existing production equipment as building blocks for new manufacturing systems [4, 5].

Reconfiguration of RMS needs to be planned ahead. New systems can rarely be realised without any engineering efforts. Typically 80-90% of a RMS can be assembled from existing modular parts [6], the rest of the modules needs to be specifically developed. The quality of the new modules is a key indicator. Questionable quality of these modules, caused by hasty engineering efforts, leads to unreliable performance when production is ramped up.

This paper presents a scientific framework, to be applied as a systems engineering tool, that consciously enables a periodic 'Zigzagging' motion between product design and manufacturing engineering. It increases mutual understanding and the effectiveness of negotiation between product- and equipment-engineers. The method is based on the definition of 'Domains' as defined by the 'Axiomatic Design' technique developed by MIT [7, 8].

2   Integral System Engineering in Product Design and Design of Reconfigurable Manufacturing Systems (RMS)

Decomposition of a product- or equipment-design and their definition at sub-levels are not performed in a single complex development effort. Instead, progression is made in successive development cycles. Moments of evaluation, to inventory residing risks in the system, are an essential part of the design loop. This was described for RMS in recent work by a graphical representation of a general development cycle as shown in figure 1 [9].

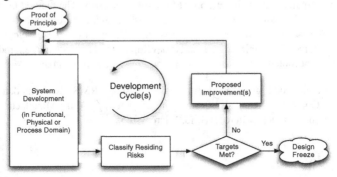

Fig. 1. General Product/Process Development Cycle. The development cycle can be used for the functional, product and/or the process domains.

The product/process development cycle describes a 'Development stage', consisting of functional decomposition and analysis at sub-levels, followed by a classification of residing risks. Depending on the outcome of the classification process a corrective action to improve the system is determined and executed. Basically all known design methodologies that have been developed in the past can be used in the development cycle e.g.; proven tools for system development are: QFD, morphological matrices, Pugh matrices. SADT/IDEF0 are typically applied for sequential processes like manufacturing. The classification of residing risks in the system can be done with Failure Mode Effect Analysis (FMEA) or Qualitative Analysis (QA). Suitable applications of the design loop have been described in [6, 9 & 10]. The method has proven to be of good use to the various system design processes but it has problems addressing the different domains concurrently. The method of Axiomatic Design is used to expand the development cycle to enable this.

The Axiomatic Design theory describes the domains and how to connect them. Typically the domains applied are; the functional domain, the physical domain and the process domain. The domains are related as shown in figure 2.

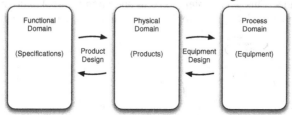

Fig. 2. Relational mapping of the domains. Specifications in the functional domain are brought into relation to design parameters of a product in the physical domain by the product design process. The design parameters are related to the process domain by the equipment design process.

At the start of a project, the functional requirements are decomposed in levels, thus dividing the project into smaller parts that can be better understood. The decomposition simultaneously takes place for the other domains to fortify the concurrent character of the development process. This process is called zigzagging [8]. Figure 3 shows the process of zigzagging in the relational map of figure 2. Zigzagging aims for a cross-domain harmonisation of the requirements, product design and equipment. The amount of remaining work is broken down and equally optimised over the domains. For the remaining project risks, a conformable breakdown should be obtained.

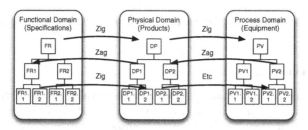

Fig. 3. While zigzagging, a hierarchical descent is made in the design process for as well specification, product design and production means. It leads to simultaneous decomposition of all domains.

However, relations between the domains should be modelled by an appropriate method of system engineering. In practice, as explained above, these methods are dependent on the domains and therefore not the same for product and equipment design. Due to this incompatibility, the process of zigzagging is not automatically congruent with the cyclic development procedure. The development cycle is not able to address all domains at the same time. Different cycles should be adjusted in some way to align the concurrent procedures and synchronise design processes in simultaneous development cycles. Since decomposition, definition at sub-levels, and risk analysis are performed in a different manner, the monitoring stage 'targets met', basically a Boolean indicator, will be the most suitable stage to compare the remaining project risks. The risk classification outcome typically is a measure that can be prioritised. This enables comparison with different risk-analysis-techniques, such as FMEA or QA. So for an improved development cycle, that is capable to support more than a single domain, the outcome of the risk classification stage is compared across the domains. This was implemented as shown in figure 4. The improved 'Multi Domain Development Cycle' was applied on an industrial case for testing.

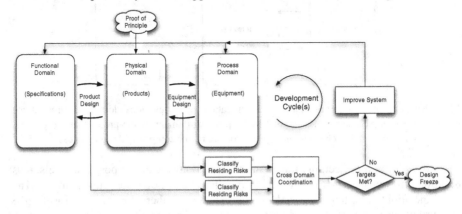

Fig. 4. Multi Domain Development Cycle for cross-domain coordination. The risk classification results from product design and equipment engineering processes are taken into general consideration. Joint classification if targets are met and what to do to fulfil them in the next cycle are point of discussion before definition of improvements. Struggles for engineers are administered and equally divided.

3   Case: Manufacturing of an Environmentally Friendly Circuit Board

3.1   Definition of the Product

The applied case, manufacturing of an environmentally friendly wireless sensor system, consists of a fairly simple electronic circuit that has to be realised in large quantities. Due to the fact that most wireless sensor systems cannot be recycled at the end of their life cycle, the aim is to minimise the environmental footprint of these sensors when disposed. This is done i.a. by replacing the standard printed circuit board with an environmentally friendly alternative (figure 5). This circuit board is embossed in a biodegradable plastic using a carbon paste to realise conductive tracks [11]. The circuit board, as required for this case, introduces a number of new features. First, the electronic layout is embossed in a plastic part, secondly, the part is filled with a carbon paste to create the conductive tracks. The electronic parts are placed in the carbon paste prior to curing. A regular pick & place process is applied.

Fig. 5. Left, printed circuit board of a wireless sensor system. Right, circuit board made of bio degradable PET.

After curing, the circuit is electrically functional and can undergo further mechanical assembly. The paper focusses on the design modifications of the plastic part in which the circuit is embossed in relation to its manufacturing process. The manufacturing process is to be implemented as a modular and reusable part of an RMS.

3.2   Application of the Multi Domain Development Cycle in a Design for Manufacturing Process

First, the state of the art PCB integration was compared to the envisioned embossed circuit board using a concept selection matrix (also known as Pugh's alternate technology matrix). Parallel to this analysis, SADT was applied on the product flow through the RMS; the most significant risks were determined. Based on the outcome of the Pugh matrix and the SADT, the cross-domain negotiations were executed and follow-up actions for the product and production development processes were defined.

This procedure was repeated in three stages in which product and process development were put in sync. The full analysis is given in figure 6. After completion, a solution was found that matched the goals of the product designers as well as the manufacturing engineers. The iterative process was cycled for three times in total. A result was found that satisfied the goals of the product designers as well as the manufacturing engineers.

4   Discussion and Conclusions

4.1   Design Information Flow when Concurrently Applying the Development Optimisation Cycle

The systems engineering cycle, as proposed in this paper, was successfully applied to the development of a new method for lens stack-alignment and its manufacturing equipment. The question arises if this could also have been the case if this method had not been applied. Processes of industrialisation for hybrid micro systems are diverse and involve large investments. This makes an objective reference measurement expensive and heterogeneous. What can be concluded is:

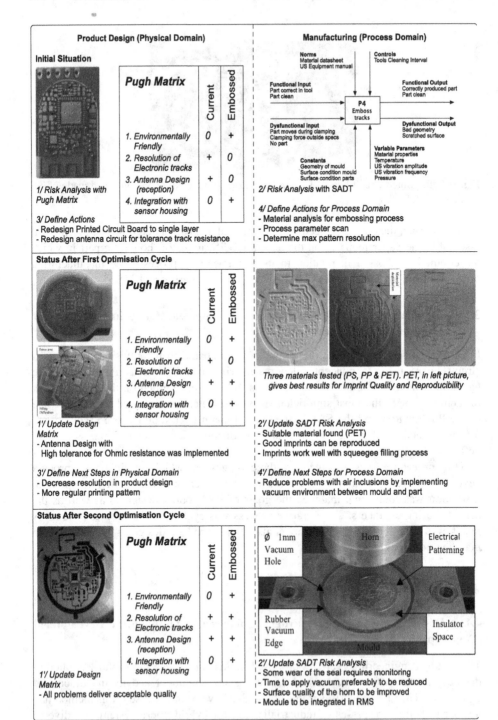

Fig. 6. The Multi Domain Development Cycle in three consecutive steps

At first, the method organises the design process of product design and manufacturing equipment. It monitors functional progression in development. This addresses development risks in an optimal order, structurally reducing the hazards of project delay. Further engineering may be considered to be 'work' instead of 'investigation' leaving minimal risks left.

Secondly, the optimisation cycle implements a zigzagging motion between the functional, physical and process domains. All domains will be decomposed hierarchically and consecutively defined at the various sublevels. This is conforming the definition of concurrent engineering in the axiomatic domain [12]. Communication between the domains is greatly enhanced as ideally proposed in the Axiomatic framework. This is shown in figure 7.

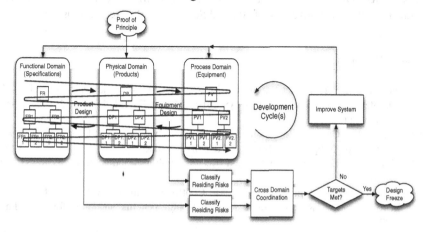

Fig. 7. Zigzagging through the domains. Decomposition and definition evolve concurrently

Thirdly, the risk inventory is updated with a regular frequency. This enables a continuously updated view on the remaining project risks. The most dominant risks can be addressed in a forehanded way, by timely prioritising the largest risks and pulling them forward in time.

Finally, the staged visualisation of project risks will elevate the level of communication in the organisation, widening the scope of personnel to be addressed. Though the engineers profit by an overview of remaining project risks, the higher level of management will also be capable of understanding the project status. This reduces discrepancy in the estimates of perceived effort to complete the project; less explanation to the management is needed.

These four effects together lead to a quicker and more structured optimisation of problems in the concurrent design of products and equipment. Visualisation of progression in development, the appropriate feedback cycle and the improved communication with technological and operational management will lead to a better architecture of product and production means at a more competitive cost.

The method as described in this paper, although applied for RMS, is truly generic and can be used for the analysis of a diversity of processes in the development and engineering stadium. The method is flexible due to the ability to fit domain specific tools for analysis in the development cycle.

4.2   Conclusion

The RMS Optimisation Cycle, combining SADT and an improvement cycle with a layered structure, can be successfully applied to monitor development of RMS. Visualisation of development-progression, the appropriate feedback loop and the improved communication with technological and operational management, will lead to a better system architecture of product and production means at a more competitive cost. The results come available in the early stage of development when the product design is not rooted yet and combined optimisations are still possible.

Acknowledgements. This research was funded by MA3 Solutions, TNO Science & Industry & the HU University of Applied Sciences in the Netherlands.

References

1. Puik, E., van Moergestel, L.: Agile Multi-parallel Micro Manufacturing Using a Grid of Equiplets. In: Ratchev, S. (ed.) IPAS 2010. IFIP AICT, vol. 315, pp. 271–282. Springer, Heidelberg (2010)
2. Koren, Y.: General RMS Characteristics; Comparison with Dedicated and Flexible Systems, vol. 3, pp. 27–45. Springer, Heidelberg (2006)
3. ElMaraghy, H.A.: Changeable and Reconfigurable Manufacturing Systems. Springer (2009)
4. Pritschow, G., Kircher, C., Kremer, M.: Control Systems for RMS and Methods of their Reconfiguration. Reconfigurable Manufacturing Systems and Transformable Factories (2006)
5. Colledani, M., Tolio, T.: A Decomposition Method to Support the Configuration / Reconfiguration of Production Systems. CIRP Annals - Manufacturing Technology 54(1), 441–444 (2005)
6. Puik, E., Smulders, E., Gerritsen, J., Huijgevoort, B.V., Ceglarek, D.: A Method for Indexing Axiomatic Independence Applied To Reconfigurable Manufacturing Systems. Presented at the 7th International Conference on Axiomatic Design ICAD, 1st edn., Worcester, vol. 7, pp. 186–194 (2013)
7. Bell, A., Gossard, D.: Suh: On an axiomatic approach to manufacturing and - Google Scholar. Journal of Engineering for Industry (1978)
8. Suh, N.P.: The principles of design. Oxford University Press, USA (1990)
9. Puik, E., Telgen, D., Moergestel, L.V., Ceglarek, D.: Qualitative Product/Process Modelling for Reconfigurable Manufacturing Systems. Presented at the International Symposium on Assembly and Manufacturing ISAM (2013)
10. Puik, E., Telgen, D., Moergestel, L.V., Ceglarek, D.: Structured Analysis of Reconfigurable Manufacturing Systems. Presented at the International Conference Flexible Automation and Intelligent Manufacturing, FAIM 2013 (2013)
11. Gielen, P., Sillen, R., Puik, E.: Low cost environmentally friendly ultrasonic embossed electronic circuit board. Presented at the Electronic System-Integration Technology Conference (ESTC), vol. 4, pp. 1–7 (2012)
12. Gonçalves-Coelho, M.A., Mourão, A.: Axiomatic design as support for decision-making in a design for manufacturing context: A case study. International Journal of Production Economics 109(1), 81–89 (2007)

The SMARTLAM 3D-I Concept: Design of Microsystems from Functional Elements Fabricated by Generative Manufacturing Technologies

Markus Dickerhof[1], Daniel Kimmig[1], Raphael Adamietz[2], Tobias Iseringhausen[2], Joel Segal[3], Nikola Vladov[3], Wilhelm Pfleging[4], and Maika Torge[5]

[1] Institute for Applied Computer Science, Karlsruhe Institute of Technology, Karlsruhe, DE
[2] Fraunhofer Institute for Process Automation, Stuttgart, DE
[3] Manufacturing Research Division, Faculty of Engineering, University of Nottingham, Nottingham, UK
[4] Institute for Applied Materials Karlsruhe Institute of Technology, Karlsruhe, DE
[5] Karlsruhe Nano Micro Facility, H.-von-Helmholtz-Platz 1, 76344 Egg.-Leopoldshafen, DE

Abstract. Generative manufacturing technologies are gaining more and more of importance as key enabling technologies in future manufacturing, especially when a flexible scalable manufacturing of small medium series of customized parts is required. The paper describes a new approach for design and manufacturing of complex three dimensional components building on a combination of additive manufacturing and e-printing technologies, where the micro component is made up of stacks of functionalized layers of polymer films. Special attention will be paid to the "3-d" modeling approach, requested to support the applicaton developer through provision of design rules for this integrated manufacturing concept . Both, the application concept as well as the related equipment and manufacturing integration currently are currently developed further in the project SMARTLAM, funded by the European Commission.

Keywords: flexible scalable manufacturing, smart manufacturing, additive manufacturing, printing technologies.

1   Introduction

Today´s fabrication methods for micro and nanotechnology enabled devices require expensive tooling and long turnaround times, making empirical, performance-based modifications to the product design expensive and time consuming. Thus till to date are often limited in their flexibility, so that complex devices, that incorporate on-board valves, membranes, discrete parts, or electrodes, cannot be developed or adapted without considerable expense in molds and assembly fixtures.

These boundary conditions create a barrier to the development of small to mediul series of complex and higher functionality devices, where the cost-benefit ratio of incorporating functionality is too risky for the typical laboratory, diagnostic or medical device developer. To bridge the gap between a high volume production with specialized equipment and a - until today - not efficient production of medium series, SME´s need to find other, more flexible and scalable approaches to produce microsystems in high volumes.

S. Ratchev (Ed.): IPAS 2014, IFIP AICT 435, pp. 147–160, 2014.

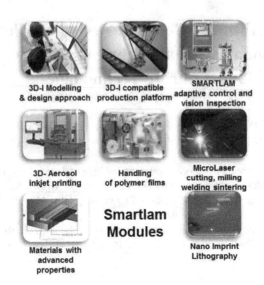

Fig. 1. SMARTLAM enabling technologies

The solution proposed by the EC-funded project SMARTLAM and presented in this paper builds on a modular, flexible, scalable scenario combining state of the art developments in technologies and materials:

- Rapid prototyping technologies in a wider sense and laminated object manufacturing (LOM) in the narrow sense - an established rapid prototyping technology building on layer by layer lamination of functionalised film sheets with different material properties is in the focus of the activities [1].
- Printing technologies, where aerosol-jet printing is in the specific focus of the project allowing for an efficient and precise manufacturing of conductive tracks, electrodes, etc.

Novel polymer film materials with advanced material properties such as anisotropic conductive film or effects arising from combinations of composite sheets will be combined with state of the art, scalable 3D printing, structuring and welding technologies as well as the usage of.

Both technologies will be integrated in a modular manufacturing environment allowing for the production of complete 3D Microsystems.

The approach proposed by SMARTLAM is designed to address the manufacturing of small medium series of micro enabled components, in contrary to the research field of roll-to-roll manufacturing focussing on a high throughput manufacturing of e-printed devices such as flexible electronics, flat panel displays or organic photovoltaics [2].

The SMARTLAM design approach builds on the assumption that most applications can be designed using modular building blocks with dedicated process sequences for each functional element – the 3 dimensional integration (3D-I) approach.

To allow for a better use of the new capabilities arising from this 3D-Integration a testbed will be set up and evaluated by two SME companies in the field of bioanalytics and lighting application. The companies are acting as potential customers, whose application requirements will be providing input to the 3D-I approach from a technical and economical perspective.

2   Application Oriented Modeling Elements of the SMARTLAM 3D-I Approach

To facilitate the development of new applications the SMARTLAM consortium introduced a modelling hierarchy allowing for structuring of the different levels of detailing and the mapping of technological capabilities, after the initial function requirements have been clarified and a first decision for a specific design has been made.

Over the first months the focus of the discussions was on the identification of an initial set of functional elements which will become expanded during the project. Sets of related process sequences for manufacturing are currently identified for each of these elements These specific process chains support the implementation of a specific functional element, representing an instance of more or less application independent manufacturing processes.

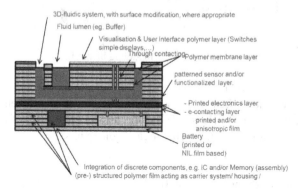

Fig. 2. SMARTLAM demonstrator for "intelligent combination of functionalized" polymer layers

2.1   Modelling of Generic SMARTLAM Functional Elements

This initial set of functional elements shall cover a broad variety of potential applications without being limited to the two main application fields, which are in the focus of SMARTLAM.

The requirements to the functional elements have to address the manufacturing methods inherent to SMARTLAM: They should either

-   have the potential to become realised by a combination of the technologies available in one level (grooves, conductive tracks,...), or

- can be manufactured by a combination of layers with different properties, e.g. a valve, realised by an elastomer layer in between two rigid layers with circular cut outs.

Fig. 2 provides an example for a (hypothetic) disposable for bio analytics device, in which fluidic and electronic functional elements had been combined by an "intelligent" combination of layers of functionalised polymers with different material properties. Besides functionalised layers, e.g. for e-contacting, layers with dedicated material properties or micro structured layers examples for integration of discrete parts are included, Positioning and contacting of a discrete chip or a battery widens the capabilities of the additive manufacturing capabilities.

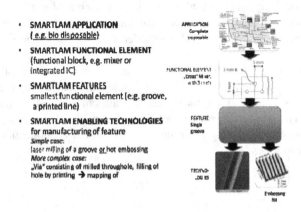

Fig. 3. SMARTLAM modeling hierarchies

In SMARTLAM the following types of functional elements for printed electronics had been identified over the first months and are currently subject to further investigations and discussions with respect to fulfill the application requirements as well as to meet the technological constraints:

Printed electronics

- electrode structures [line, interdigital]
- MR Sensors
- Capacitive switches
- Functional elements for the energy storage
- Printed batteries
- NIL-based Batteries
- Components for electronics
- Resistors
- Switches

Functional elements for micro fluidics

The concepts allows for a broad range of fluidic functional elements to be investigated in a more detail, especially in M7-12.

- Microfluidic channels
- Membranes
- Valves
- Mixers
- Storage
- Actuators

Functional elements building on surface modification

SMARTLAM technologies basically allow for an active treatment of surface properties (chemical and physical properties). Surface activation (e.g. Ionisation) or an active control of wttability by laser processing are just two examples to mention in this context.

Composite functional elements

Some of the functional elements can be combined to composite functional elements of a "higher" integration level. The functional element "positioning and integration of a chip" may consist of two functional elements "milling of a pocket" and "e-contacting"

Implementation strategies

Many of the functional elements mentioned above can be realized in a single manufacturing step, where the function can be realized by manufacturing of a single geometric element, which will be covered in more detail in the next paragraph. In most cases even more complex functional elements can be described as a combination of such features. From a process perspective however, there are typically multiple solution strategies for implementing functional elements. The "Via" functional element may serve as an example, as technological as well as chemical solutions are feasible and will have to be selected depending on the application boundary conditions. These solutions include but are not limited to:

- realisation of a "mechanical" contacting between film layers consisting of a through hole filled with electronic ink.
- Physical realisation of a contacting using the special material properties e.g. of "anisotropic conductive films"

2.2   Features

"Features" represent a kind of intermediate between the application-oriented function element and the "manufacturing output", mostly resulting in a set of geometric primitives that can be micro milled or cut, coated, printed, welded, etc.. The following examples may illustrate underlying concept:

- A single hole can be a vertical microfluidic channelas well as a pocket for integration of discrete parts.
- An array of such micro might represent a "micro sieve" for blood separation in the bio disposable application areas.

- A combination of the "hole"-feature with other features such as the filling of a hole with e-ink can result in a contacting for manufacturing of "vias", know from printed circuit board technology

Typical features:

- Channels (for fluidic or optical properties)
- Pockets (cavities, lumen or locating holes)
- Printed lines (conductive tracks, sensors, "via")
- Material layers with properties, different from polymers (realisation of batteries, realisation of membranes (elastomers)

Evaluation tests could demonstrate the validity of this approach and similar concepts on feature level had been successfully tested in other micro related contexts [3]. A thorough evaluation in order to validate the principles of this concept will be performed over the course of the project.

3   Selection of Manufacturing Process Chain and Product Design

An integrated micro device often consists of a very large number of features which form the application oriented functional elements. For the fabrication of such devices a manufacturing process chain of high complexity is required. To address this problem, in SMARTLAM the product design and the process selection are incorporated into a hierarchical model where decisions are made at different abstraction layers. Similar hierarchical approaches have been elaborated and realised in CIM or STEP [4][5].

3.1   Product Design Method

In the SMARTLAM 3D-I concept the product designer uses a library of structural features which can be integrated to built functional elements. These features together with information about technologies that can be used for their fabrication are stored in a database. In the hierarchical model the first layer is the library and is the only layer open to the designer. The second layer is the pool of available technologies and forms the manufacturing process chain. The third layer is the process parameters and is used for the optimisation and selection of specific process chain (see Figure 4). If the simple example from the previous section is taken, the library will offer a variety of micro holes in a range of sizes and materials which can be arranged to form a micro sieve. Some of the technologies that can be used for the fabrication of wholes in a polymer material are micro milling, laser machining or micro- injection moulding. The process parameters for the machining of the specific holes selected by the designer are stored in the database. After the micro sieve the next functional element on the device might be a conductive line. Possible technologies for its execution would be any kind of lithography, aerosol or jet printing. The design can be continued with the next device feature or a lamination step. Thus is it is apparent that an effective method of selecting the best available chain has to be created. The advantage of using a library of preset geometries is that the designer is limited to structures which fabrication is already a well-established process. The number of alternative process depends on the completeness of the database which can be expanded when required. The addition of each new device feature multiplies the number of the possible manufacturing process chains by a factor equal to the number of available technologies for the feature. Therefore the development of fast and effective way of selecting the optimum chain is essential.

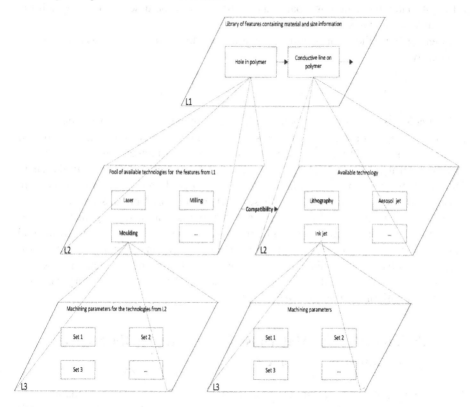

Fig. 4. SMARTLAM process sequence hierarchies

3.2   Process Chain Selection Method

Large number of the initially identified process chains are rejected by breaking some of the links on the second hierarchical level. This can be done either on material-technology (M-T) or technology- technology (T-T) compatibility criteria. (see Figure 4). The T-T compatibility criterion is relatively simple and normally there is a clear rejection or approval of a specific chain. In many cases this is the inability of combining vacuum with non- vacuum technologies. As an example, laser machining cannot be combined with electron lithography due to the extreme complications of the workpiece transfer. In cases when the technology compatibility assessment is not that straightforward SMARTLAM can exploit the Process Pair Interface Model developed under the EC FP7 funded EUMINAfab project which provides more detailed analysis of the pair maturity level [6]. The M-T criterion is more complicated and requires some more detailed investigation. For many technologies there is a "favourite" material but this does not exclude the use of some other materials. For example PET is one of the most popular substrate materials for aerosol jet printing but shows poor UV laser machinability. This requires the development of a scoring system for different material- technology pairs and setting of threshold score below which the manufacturing chain is rejected.

After applying the compatibility criteria the remaining process chains are arranged in an array:

$$Array = (chain_1 \quad ... \quad chain_{i-1} \quad chain_i \quad chain_{i+1} \quad \cdots \quad chain_n)$$

Technologies that can be associated with more than one set of process parameters are linked in a separate chain for each set. The minimisation of the array is executed in two phases. In the first one process parameters such as temperature or pressure are assessed and it is checked whether any of the technologies or materials in the chain imposes restrictions to these parameters. Chains which do not meet these criteria are rejected. The second minimisation phase is when the actual process optimisation is performed. This is done by comparing the process parameters against the user requirements. User requirements can be machining time, cost or accuracy. At the end only the best matching chain remains which is used for the fabrication of the designed device. An advantage of this approach is that the selection of the technologies also serves for identification of the process parameters. This allows for very fast and efficient reconfiguration of the SMARTLAM modules.

4   Technologies and Materials of Relevance for the Smart Manufacturing Approach

Representing the smallest building blocks in modelling of process chains three different types of technologies are currently under evaluation regarding their integration in the system setup:
- Technologies for additive manufacturing and e-printing aerosol jet printing. In SMARTLAM the e-printing functionality is realized by aerosol jet printing, offering good results for manufacturing of line-based geometries such as conductive paths [8]
- Handling assembly and bonding technologies [7]
- Technologies for direct and indirect milling and cutting of polymer films [laser milling, cutting, nano imprint lithography, hot embossing]

Each of these technical sections will be briefly introduced with respect to its specific relevance for the Smartlam approach.

4.1   Structuring Technologies

For (pre-) structuring of films, technologies such as laser direct structuring but also indirect technologies such as NIL or hot embossing are under evaluation. Criteria to be mentioned in this context are scalability of processes, machine unit costs, required precision, small lot sizes vs. larger volumes.

First tests have been conducted at KIT-IAM to evaluate typical SMARTLAM features and functional elements. In SMARTLAM polymers such as polycarbonate (PC) or poly(methyl methacrylate) (PMMA) are used. Those polymers show a high laser beam absorption in the ultraviolet (UV) (248 nm, 193 nm) as well as in the infrared (IR) (10.6μm). From previous studies it is known that a high optical absorption at UV wavelengths leads to low material removal rate, which in turn enables a precise control of the achieved ablation depth and a good surface quality [9]. With UV laser structuring lateral resolutions in the sub-micrometer range via direct writing or direct optical imaging of complex mask structures can be achieved. Furthermore below or in the range of the laser ablation threshold of the polymer a surface modification is possible. With optimized laser and process parameters surface modification can be applied in order to adjust the wettability of polymer surfaces from superhydrophilic to superhydrophobic properties.

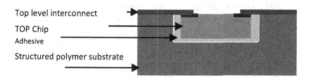

Fig. 5. Top Level interconnect design for Chip integration

IR laser structuring of polymers can provide a very flexible material processing with high processing speeds. High accuracy channel widths down to 50 μm can be produced in PMMA. Nevertheless, due to the fact that IR laser-assisted ablation processing is mainly thermally driven, a main challenge is to achieve small structures (<50 μm) with minimized heat affected zone.

An appropriate combination of UV and IR laser processing can be applied for the generation of functionalized polymeric micro devices [10]. Fig. 6 and Fig. 7 show laser structured thin polymer films using IR and UV laser systems.

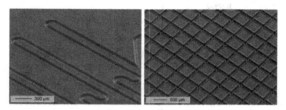

Fig. 6. Micro channel structures in PMMA achieved by IR laser ablation (wavelength: 10.6 μm)

Fig. 7. Pocket structure in PET achieved by UV excimer laser ablation (wavelength 193 nm)

4.2   Polymer Film Handling and Assembly Related Technologies

The polymer film handling addresses a limiting aspect with respect to the achievable precision: the handling and (fine) positioning of parts within the systems. Other technologies such as dispensing and bonding of polymer films are subject to a detailed analysis as well. While there exists a long experience in handling of polymer films the proper alignment of the sheet layers with respect to the required accuracy of a few μm is still challenging. Manufacturers have to deal with selective thermobonding to address the relaxation behaviour of the film material without loss in accuracy.

The polymer film handling process is shown in Fig. 8. The first step, feeding, supplies singular film sheets. This is challenging, as the films require to be provided as flat as possible for the precision pick and place process and have to be "loose" in order for the gripper to grab it. A magazine with an air suction for each sheet could be a possible approach.

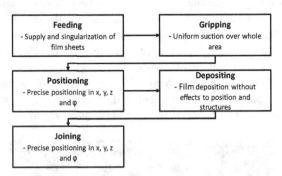

Fig. 8. Polymer film handling process

The second process step is gripping. As the film sheets are flexible, but still have to be precisely positioned, a uniform suction over the whole sheet area is pursued. Deformations during the gripping process have to be avoided at any cost, as these would lead to major deviations between the respective product layers. A flat vacuum gripper with an arrangement of holes <<1mm could be a possible approach.

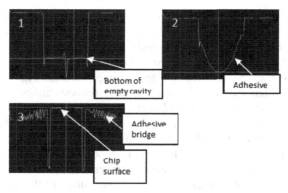

Fig. 9. Stages of chip assembly

The third step is the positioning of the sheet relative to the substrate. This requires positioning in position (x, y, z) and orientation (φ). The required precision depends on the type of layer. If the sheet features micro-structures, which require relative alignment to the substrate, the required positioning accuracy might be <25μm. If the sheet does not feature any micro-structures, the positioning accuracy is rather uncritical. A cartesian precision placement system with an additional rotational axis is proposed here.

Letting the sheet go is as critical and challenging as the gripping process. While it is very important to attach the sheet as uniformly as possible without any deformation, no deformation may happen during the disengagement. At the same time neither substrate nor sheet may be damaged during disengagement. A possible strategy could be a light blow-off of the sheet.

For the joining process several options exist, such as adhesive bonding, induction welding, laser welding, lamination or thermobonding [14]. The requirements to be addressed are joining of the sheet with the substrate without deformation, damage and voiding. Currently, a lamination process is investigated, as it seems to fit very well to the requirement profile.

For the integration of the functional elements "integration of discrete part" and the related technologies "milling of a pocket", "dispensing of glue" and "positioning of chip" first tests had been conducted to evaluate the feasibility.

The "milling of a pocket" is done by application of the above-mentioned structuring technologies. The further technologies are merged to a discrete part assembly process. For the functional element "integration of discrete part" first tests have been carried out with a top level interconnect design for chip integration (see Fig. 5.)

The chips have to be fixed in the cavities with adhesive bonding. In addition, the dispensed adhesive has to fulfill another important function: It works as a convex or concave bridge filling the gap between the LED and the substrate, on which electrical circuits can be printed. Furthermore the adhesive has to protect the chip to ensure reliability during operation. The length of the chips edge is smaller than 500 μm. The corresponding pockets are slightly larger. Therefore the amount of adhesive is in the low nanolitre range and has thus to be dispensed very precisely. Furthermore adhesive forces between the walls of the pocket and the adhesive disturb the accurate and repeatable positioning of the droplet. To still achieve a stable assembly process, high precision measuring equipment for the lateral and vertical dimensions of the pockets and the dispensed adhesive as well as a precision placement system with a positioning accuracy in the low micron range for the dispensing tool is required. The same high requirements have to be fulfilled for the chip placement. Fig. 9 shows optical measurements of three stages of the chip assembly process: at first the empty pocket is filled with adhesive and the assembled chips with the adhesive bridge.Afterwards the adhesive needs to be cured. The products assembled within SMARTLAM are based on polymer films. Currently a broad range of substrate materials is intended to be applied, e. g. polyethylene terephthalate (PET), polycarbonate (PC) or poly(methyl methacrylate) (PMMA). The glass transition temperatures (PET: 70 °C; PC: 145 °C; PMMA: 105 °C) should not be exceeded during the curing of the adhesive. For most common adhesives comparative high temperatures above 100 °C are needed for fast curing. To maintain low cycle times with low temperatures UV curing adhesives are taken into account. Laser as part of the SMARTLAM system could potentially be used for precise and local curing to reduce the heat-affected zone compared to a convection oven.

The SMARTLAM e-printing functionality is used in a following step for the electrical connection of the chips.

4.3   Materials Properties of Specific Relevance for SMARTLAM Concept

Novel polymer film with advanced material properties are of a specific interest for the SMARTLAM approach. There exists a broad range of material features that will become of interest for SMARTLAM applications:

"Standard materials"

The use of polymer materials with dedicated optical and mechanical properties provides opportunities for a broad range of applications but does also cause problems while its application in the SMARTLAM context. While the activities in the startup phase are focussing on polymer films, the application of ceramic films, similar to LTCC applications [15] as well as the application of flexible glass [16] but also combinations of different substrate layers with surface properties will be subject to the future research.

In addition materials with advanced chemical and physical properties will be evaluated in later stage of the project. As an outlook may serve the following materials:

- Anisotropic conductive films allowing for the an electrical conductivity vertical to the sheet plane, While the application in LCD panel production and chip industry is well established the application of the ACF in flexible environment is still subject to research activities, e.g. in the field of flip-chip on flex packages assembly [11]
- high optical transparency, robust flexibility, and excellent conductivity caused by general synthesis of aligned carbon nanotube/polymer composite films with. (potential applications such as flexible conductors for optoelectronic devices) [12]
- Actuation of the microsystem, caused by electro active properties of the polymer [13]

4.4   System Integration on Process Chain Level

According to the 3D-integration paradigm different combinations of process sequences –each representing the respective process sequences for manufacturing of the respective design building blocks. Figure 10 provides an example how such process sequences could look like

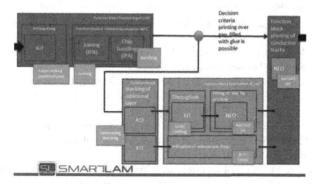

Fig. 10. Example of a process sequence for connecting a discrete part

Different setups are currently under evaluation and testing, contributing to the decision support tools as well as the database.

5   Conclusions and Outlook

In the paper a novel approach for flexible scalable manufacturing of micro components building on a combination of laminated objects modelling, integration of laser technologies for structuring of films and printed electronics for e-contacting and printing of electronic components was presented.

Specific attention was paid to the design of applications, building on a new approach for 3D-integration, (3D-I). Design elements on different aggregation levels and their implementation have been introduced.

Over the next month these concepts will be developed further and adapted to the needs of the two project demonstrators in the field of lighting and microfluidics

Acknowledgements. The research related to the SMARTLAM flexible scalable manufacturing approach receives *funding* from the European Community's Seventh Framework Programme (FP7/2007-2013) Grant agreement No. 314580.

Finally, the support by the Karlsruhe Nano Micro Facility (KNMF, www.kit.edu/knmf) for laser processing is gratefully acknowledged.

References

[1] Muellera, B., Kochan, D.: Laminated object manufacturing for rapid tooling and patternmaking in foundry industry. Computers in Industry 39(1), 47–53 (1999)

[2] Schwartz, E.: Cornell University MSE (2006),
http://people.ccmr.cornell.edu/~cober/mse542/page2/files/Sch
wartz%20R2R%20Processing.pdf

[3] Albers, A., Börsting, P., Wildermuth, P., Haußelt, J., Kraft, O.: In: Karlsruhe Kraft, O., Haug, A., Vollertsen, F., Büttgenbach, S. (hrsg.) Kolloquium Mikroproduktion und Abschlusskolloquium SFB 499, Oktober 11-12. KIT Scientific Reports; 7591. KIT Scientific Publishing (2011)

[4] Xu, X., Nee, A.Y.C.: Advanced Design and Manufacturing Based on Step. Springer (2009)

[5] Dickerhof, M., Mampel, U., DIdic, M.: Workflow and CIMOSA, A background case study. Computers in Industry 40, 197–205 (1999)

[6] He, X., Gao, F., Tu, G., Hasko, D., Huettner, S., Steiner, U., Greenham, N.C., Friend, R.H., Huck, W.T.S.: Formation of Nanopatterned Polymer Blends in Photovoltaic Devices. Nano Lett. 10, 1302–1307 (2010)

[7] He, X., Gao, F., Tu, G., Hasko, D., Huettner, S., Steiner, U., Greenham, N.C., Friend, R.H., Huck, W.T.S.: Formation of Nanopatterned Polymer Blends in Photovoltaic Devices. Nano Lett. 10, 1302–1307 (2010)

[8] King, B., Renn, M.: Aerosol jet direct write printing for mil aero electronic applications: Source: Optomec, Inc., 3911 Singer, NE, Albuquerque, NM 87109, USA

[9] Pfleging, W., Kohler, R., Südmeyer, I., Rohde, M.: Laser Micro and Nano Processing of Metals, Ceramics, and Polymers. In: Laser-Assisted Fabrication of Materials. Springer Series in Materials Science, vol. 161, pp. 319–374 (2013)

[10] Pfleging, W., Kohler, R., Schierjott, P., Hoffmann, W.: Laser patterning and packaging of CCD-CE-Chips made of PMMA. Sensors and Actuators B: Chemical 138(1), 336–343 (2009)

[11] Chan, Y.C., Luk, D.Y.: Effects of bonding parameters on the reliability performance of anisotropic conductive adhesive interconnects for flip-chip-on-flex packages assembly II. Different bonding pressure. Microelectronics Reliability 42(8), 1195–1204 (2002)

[12] Peng, H.: Aligned Carbon Nanotube/Polymer Composite Films with Robust Flexibility, High Transparency, and Excellent Conductivity Division of Materials Physics and Applications, Los Alamos National Laboratory, Los Alamos, New Mexico 87545. J. Am. Chem. Soc. 130(1), 42–43 (2008), doi:10.1021/ja078267m

[13] Bar-Cohen, Y.: Electroactive Polymer (EAP) Actuators as Artificial Muscles: Reality, Potential, and Challenges, SPIE ebooks, 2nd edn. (2004),
http://ebooks.spiedigitallibrary.org/book.aspx?bookid=146
(April 5, 2013)

[14] Tahhan, I.: Ein Beitrag zum wirtschaftlichen Fügen von mikrofluidischen Baugruppen, Universitätsbibliothek Freiburg, Dissertation (2009)

[15] http://www.schott.com/epackaging/german/download/schott_db_l
tcc_via_rz_d_2012_03_29indd.pdf (April 6, 2013)

[16] http://www.convertingquarterly.com/Portals/1/files/matteucci
-awards/2011-Flexible-Glass-Substrates-for-R2R-
Manufacturing.pdf (April 6, 2013)

Optimal Design of Remote Center Compliance Devices of Rotational Symmetry

Yang Liu and Michael Yu Wang[*]

Department of Mechanical and Automation Engineering, The Chinese University of Hong Kong, Shatin, N.T., Hong Kong SAR, China
{yliu2,yuwang}@mae.cuhk.edu.hk

Abstract. Remote Center Compliance (RCC) devices are passive devices used in automated assembly. For round peg-in-hole insertions, initial misalignments of the peg can cause heavy radial loadings on the corresponding hole in arbitrary direction. RCC devices with rotational symmetry property are therefore desirable. This paper discusses how such devices may be designed and shows that circular periodic structures satisfy this property. The elastic part of the RCC device is formulated as a compliant mechanism and a systematic design methodology is proposed based on structural optimization. A smooth optimal design is achieved with distributed compliance and numerical simulation is conducted to illustrate the feasibility. The proposed method is expected to be used to design a novel device for assembling fragile plastic parts, which is a challenge to 3C industry.

Keywords: Remote Center Compliance (RCC), rotational symmetry, circular periodic structure, structural optimization.

1 Introduction

RCC devices were first developed at Draper Laboratory by Whitney and others [1-4]. Drake first pointed out that RCC devices could be created using rubber-metal sandwiches called elastomer shear pads (ESP) [3]. However, the analysis model for ESPs is inaccurate and it depends on the configuration parameters [5]. Havlik used three elastic rods to construct the device and proposed an algorithm to find rod parameters in [6]. Ciblak also investigated the possibility of using prismatic beams [7]. Drake and Hricko designed some devices conprising compliant linkages and elastic joints [2, 8].

The elastic part of an RCC device can be modelled as a compliant mechanism. As lumped compliant mechanism will induce stress concentration, distributed compliant mechanism is desirable in actual applications. The kinetoelastic model which belongs to the continuum structural optimization method is an efficient tool of designing distributed compliance mechanisms and has been utilized here [9]. Rather than a fixed topology or shape configuration, structural optimization can significantly enlarge the design domain. Additionally, the optimization model could be applied to various

---

[*] Corresponding author.

S. Ratchev (Ed.): IPAS 2014, IFIP AICT 435, pp. 161–169, 2014.

applications, such as assembly of small fragile parts in 3C industry, etc. It is therefore helpful to regard the design of an RCC device as question of structural optimization.

In the field of automated assembly, the task of fitting a round peg into a round hole is fundamental. As mentioned in the abstract, it is better to design a device with rotational symmetry property. The feasible structures that satisfy this property are discussed below. It has been proved that a circular periodic structure with at least 3 elements is workable and more elements are required if considering high-order geometric nonlinearities. The existing RCC device designs agree well with the above conclusions as most of them are circular symmetric structures with three or four elements.

The remainder of this paper is organized as follows: in Section 2, the compliant mechanism is analyzed and modeled using kinetoelastic model. The rotational symmetry property is introduced and feasible structures achieving this property are discussed. We show a schematic proof of the feasibility of the circular periodic structure. Section 3 presents formulation of the optimization model which includes the objective, constraints, design parameterization and solution scheme. The optimal design is presented and simulated in Section 4 to verify the proposed method. The conclusions and future works are discussed in Section 5.

2   Design of RCC Devices Compliant Structure

2.1   Analysis of Compliant Mechanism

Conventional and modified part mating events are shown in Fig. 1 and the shaded region is the design domain which can be modelled as a compliant mechanism. The second representation is adapted because it is easier to handle the structural optimization problem of compliant mechanism with a fixed upper boundary.

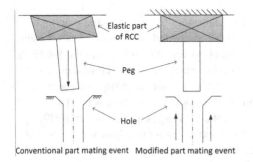
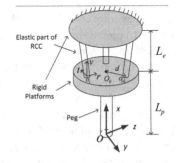

Fig. 1. Part mating events   **Fig. 2.** RCC Structure: placement and parameters

The compliant mechanism is formulated by the kinetoelastic model which concerns both kinematic motion and stiffness characteristics of compliant mechanism. According to Hooke's law, the force-displacement relationship at the rigid peg tip O under global coordinate $O-X-Y-Z$ can be expressed as

$$K_M \cdot \Delta X = P \qquad (1)$$

where K_M is the mechanism stiffness matrix, $P^T = [P_x\ P_y\ P_z\ M_x\ M_y\ M_z]$ and $\Delta X^T = [\Delta x\ \Delta y\ \Delta z\ \Delta\theta_x\ \Delta\theta_y\ \Delta\theta_z]$ are generalized force and motion at O respectively. Notably, linear Hooke's law is satisfied due to the small clearance between peg and hole.

2.2   Rotational Symmetry Design

The rotational symmetry property is introduced in the abstract. Two kinds of feasible structures which satisfy this property are shown in Fig. 3. Note that revolutionary symmetric structure is occlusive. It costs much more in materials and its stiffness at radial direction would be exceptionally higher. Therefore, circular periodic structure is more reasonable.

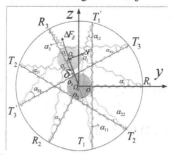

Fig. 3. Realization of rotational symmetry property: (a) revolutionary symmetric structure (only half is shown); (b) circular periodic structure

The schematic proof of circular periodic structure with N elements can satisfy this property is described in the followings. Intermittently placed springs, shown in Fig. 4, can be utilized to represent the compliance elements. Each element comprises a radial spring (spring constant: k) and two tangential springs (spring constant: k_t). A small deformation δ in arbitrary direction θ is given at O_0.

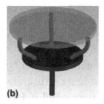

Fig. 4. Plan view of circular periodic structure (only three compliant elements are shown)

The resultant forces and directional stiffness in the direction of and perpendicular to δ together with the rotational moment and resultant stiffness can be obtained as

$$\begin{cases} \Delta F_\delta = \Delta F_{\delta r} + \Delta F_{\delta t1} + \Delta F_{\delta t2} = N(k + 2k_t)\delta/2 \\ \Delta F_\perp = \Delta F_{\perp r} + \Delta F_{\perp t1} + \Delta F_{\perp t2} = 0 \\ M_r = M_{t1} = M_{t2} = 0 \end{cases} \qquad (2)$$

$$\begin{cases} K_\delta = \Delta F_\delta/\delta = N(k + 2k_t)/2 \\ \qquad K_\perp = K_{\Delta\theta_x} = 0 \end{cases} \tag{3}$$

Combining the above results, the system stiffness K is derived as

$$K = N(k + 2k_t)/2 \tag{4}$$

which is not related to the direction θ. This completes the proof. Notably, the following trigonometric identities hold

$$\begin{cases} \sum_{i=1}^{N} \cos(\theta - \beta_i) = 0; \sum_{i=1}^{N} \sin(\theta - \beta_i) = 0; (N \geq 2) \\ \sum_{i=1}^{N} \cos[2(\theta - \beta_i)] = 0; \sum_{i=1}^{N} \sin[2(\theta - \beta_i)] = 0; (N \geq 3) \end{cases} \tag{5}$$

where $\beta_i = 2\pi(i - 1)/N$. If second-order geometrical nonlinearity is considered, the following trigonometric identities should be utilized in the proof.

$$\begin{cases} \sum_{i=1}^{N}[\cos(\theta - \beta_i) \cdot \sin^2(\theta - \beta_i)] = 0; (N \geq 4) \\ \qquad \sum_{i=1}^{N} \sin^3(\theta - \beta_i) = 0; (N \geq 4) \end{cases} \tag{6}$$

Several interesting conclusions can be drawn about circular periodic structure and are summarized as: (*i*) at least 3 discrete compliant elements are required; (*ii*)at least N complaint elements are required if $(N - 2)^{th}(N \geq 4)$ order geometric nonlinearity is considered; (*iii*)when the number of complaint elements goes to infinity, the circular periodic structure becomes a revolutionary symmetric structure.

According to Hooke's law, the force-displacement relationship at beam tip O_i under local coordinate $O_i - v - r - t$ is

$$K_E \cdot \Delta X_E = P_E \tag{7}$$

where K_E is element stiffness matrix, P_E and ΔX_E are generalized force and motion respectively. Notably, beam finite element is used to calculate K_E. With transformations of load and displacement, the mechanism stiffness matrix K_M is achieved as

$$K_M = \sum_{i=1}^{N} T_i^O \cdot K_E \cdot T_i^{O'} \tag{8}$$

where T_i^O is the transformation matrix between local and global coordinates.

Suppose the distance between peg tip and compliance center is a_x. The mechanism stiffness matrix becomes diagonal at the compliance center, and a_x can be acquired as

$$a_x = -K_M(3,5)/K_M(3,3) = K_M(2,6)/K_M(2,2) \tag{9}$$

Notably, the two ratios in Eq. 9 are equivalent according to Eq. 8.

3   Structural Optimization of RCC

3.1   Objective Function

One practical optimization criterion is to maximize the output displacement u_o which is the displacement of the contact point along the chamfer for unit normal contact force as shown in Fig. 5. According to geometric relationship equations, Eq. 1 and Eq. 8, the objective function can be expressed as

$$u_o = \Delta y / \cos \alpha = f_1(\mu) = A - \mu B \qquad (10)$$

where Δy is the displacement of the peg tip, α is the angle of chamfer, μ is the coefficient of friction, A and B (positive) are constant coefficients. According to Eq. 10, u_o is a decreasing function of μ which means that RCC device is more difficult to work under conditions of a higher coefficient of friction. Maximizing u_o is equivalent to maintaining the same displacement at the output port under conditions of a higher coefficient of friction. Therefore, the objective function is equivalent to maximizing the range of friction coefficient that the RCC device can handle.

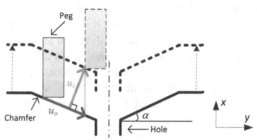

Fig. 5. Illustration of the chamfer crossing

3.2   Design Constraints

Successful Assembly Conditions. In order to cross the chamfer and avoiding wedging, the initial misalignments (y_{ho}, θ_{ho}) of the hole should satisfy the followings [4]:

$$|y_{ho}| \leq W; |\theta_{ho}| \leq (D_h - d_p)/(\mu d_p) \qquad (11)$$

where W is width of chamfer, D_h and d_p are the diameters of hole and peg respectively. In order to avoid jamming, the applied forces must satisfy the followings [4]:

$$\begin{cases} -1/\mu \leq P_y/P_x \leq 1/\mu \\ -\lambda \leq M_z/(r_p P_x) + \mu(\lambda + 1)P_y/P_x \leq \lambda \end{cases} \qquad (12)$$

where $\lambda = l/(2\mu r_p)$, r_p is the radius of peg and l is the insertion depth. When λ is small, jamming is most likely to happen and the slope of sides of the jamming diagram approaches $-\mu$. Thus, if the slope of applied force $-a_x/r_p$ is approximately equal to $-\mu$, then the applied forces and moments have the best chance of lying inside the jamming avoiding region [1]. Jamming constraint therefore can be converted into:

$$0 \leq a_x \leq \mu r_p \tag{13}$$

Constraint on Input Displacement. This is introduced to indirectly control the maximum stress levels in the compliant mechanism. For detailed explanations, please refer to reference [10]. The constraint of the input displacement u_i is expressed as

$$u_i = \Delta y \sin \alpha + \left(\Delta x - r_p \Delta \theta_z\right) \cos \alpha \leq [u_i]_{max} \tag{14}$$

where u_i is the displacement of the contact point along chamfer's normal direction, which is shown in Fig. 5.

Constraint on Coupling Stiffness. Many coupling stiffness elements will vanish in K_M due to periodic property. A large deformation would occur if the misalignment compensation capacity were increased. Then the elements in stiffness matrix will vary from each other and the coupling stiffness elements in K_M will be non-zero which means the compliance center would drift away. It order to reduce its negative effect, the element of rotational stiffness about radial direction should be reduced to make the element like a spring with no coupling stiffness. From the perspective of engineering application, the ratio of two terms less than 0.1 can be regarded as small one and the numerator term can be ignored. We therefore introduce a constraint concerning the ratio of rotational stiffness about radial direction over tangential direction as

$$S.R. = K_E(5,5)/K_E(6,6) \leq 0.1 \tag{15}$$

Constraint on Buckling and Stability. In order to maintain the stability of the compliant element, its critical buckling load should be greater than a given minimum value.

$$P_{cr} \geq [P_{cr}]_{min} \tag{16}$$

3.3   Model Formulation

In conclusion, the optimization model can be summarized as follows.

$$\begin{cases} Minimize: J = -u_o \\ Subject\ to: \begin{cases} |y_{ho}| \leq W \\ |\theta_{ho}| \leq (D_h - d_p)/(\mu d_p) \\ 0 \leq a_x \leq \mu r_p \\ u_i \leq [u_i]_{max} \\ S.R. \leq 0.1 \\ P_{cr} > [P_{cr}]_{min} \\ K_M \cdot \Delta X = P \end{cases} \end{cases} \tag{17}$$

The design variables are the distance $d = \overline{O_0O_t}$, shape Ω of compliant element cross section and neutral axis. The shape Ω at each vertical level is defined by a set of radial parameters which start from neutral axis to the surface. In this paper, two second order parabolic curves are employed to represent the neutral axis. The method of moving asymptotes (MMA) [11] is employed to solve the proposed optimization problem.

4   RCC Device Design (Numerical Example)

The design domain is defined as a cylinder with radius $R=100$ mm and height $L_e=50$ mm. The length of the peg and diameters of peg and hole are 100 mm, 12.000 mm and 12.033 mm respectively. The horizontal length and angle of chamfer are 3 mm and $\pi/4$. The initial misalignments are given as -3 mm and 0.0275 rad. The maximum input displacement is 0.2 mm and the minimum critical load is 1 KN. The Young's modulus, Poisson's ratio and coefficient of friction are 210 GPa, 0.3, and 0.1 respectively. In this paper, the RCC device with three compliant elements is designed to verify the proposed methodology. However, it is a straightforward matter to extrapolate to cover situations involving more than three elements. Note that the designed results are based on unit normal contact force.

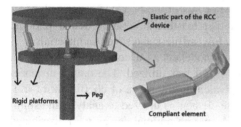

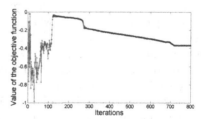

Fig. 6. Optimal design   **Fig. 7.** Iteration history of objective function

The optimal design is illustrated in Fig. 6. The optimal output and input displacements are 0.369 mm and 0.190 mm respectively. The position of the compliance center, the stiffness ratio and critical buckling load of optimal design are 0.138 mm, 0.054, and 1.682 KN respectively. This clearly shows that all the constraints are satisfied. The optimal design is modelled and visualized with LOFT operation in Solid-Works by connecting all the optimal cross sections at different vertical levels. The optimal compliance element is quite smooth and flat mainly due to the constraint of coupling stiffness. Point flexures are avoided and distributed compliance is achieved by the proposed methodology. The convergence history of objective is illustrated in Fig. 7 which shows a good convergence. The present methodology generalizes some existing designs as the realization domain of compliant element is extended from round rod with elastic joints and round beam to beam with arbitrary cross sections.

The optimal design was analyzed and simulated by Autodesk Algor Simulation. The upper boundary of compliant mechanism is connected to a fixed platform and the lower boundary is joined to a very rigid platform. The property of remote compliance center is demonstrated in Fig. 8 where pure force results in a pure translation and pure moment causes a pure rotation. Numerical errors in translation and rotation are due to the small compliance in the platform and peg.

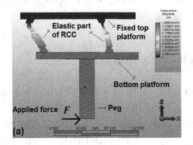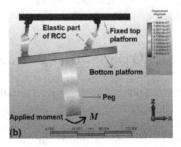

Fig. 8. Demonstration of compliance center: (a) pure translation, (b) pure rotation

5   Conclusion

We propose a novel design methodology by using structural optimization for design of an RCC device. The rotational symmetry property is discussed and it has been shown that a circular periodic structure satisfies this property. Comparing with revolutionary symmetric structure which is another feasible structure, circular periodic structure is adopted because it achieves quite low stiffness and saves on materials.

The elastic part of the RCC device was modelled as a compliant mechanism. To distribute the compliance, kinetoelastic model was utilized. The output displacement was selected as the objective and equivalent to the range of coefficient of friction that RCC device can handle. Beam finite element is utilized to analyze and simulate the complexity of design deformation. A smooth optimal design with distributed that is cost effective to manufacture is obtained in this paper. It satisfies all constraints and the properties of remote compliance center are demonstrated through the simulation.

The work reported here is far from complete. It is rewarding to integrate structural optimization with reliability constraint by considering the geometrical uncertainties of peg and hole, uncertainty of the coefficient of friction, etc. In addition, the discussed methodology can be applied to specified applications by introducing corresponding objective and constraints. For example, the applied force and vertical stiffness is critical in assembling fragile plastic parts which are easily damaged.

Acknowledgments. The authors would like to thank Prof. X. M. Wang, Prof. Y. J. Luo, Dr. L. Li and Dr. X. F. Tian for their informative discussions and comments, as well as Prof. K. Svanberg for providing his MMA codes.

References

1. Whitney, D.: Mechanical Assemblies: Their Design, Manufacture, and Role in Product Development. Oxford University Press, New York (2004)
2. Drake, S.H.: Using Compliance in Lieu of Sensory Feedback for Automatic Assembly. Ph.D. Dissertation, MIT (1977)
3. Whitney, D., Rourke, J.M.: Mechanical Behavior and Equations for Elastomer Shear Pad Remote Center of Compliance. ASME J. Dyn. Syst. Meas. Control. 108, 223–232 (1986)
4. Whitney, D.: Quasi-static Assembly of Compliantly Supported Rigid Parts. ASME J. Dyn. Syst. Meas. Control. 104, 65–77 (1982)
5. Joo, S., Waki, H., Mayazaki, M.: On the Mechanics of Elastomer Shear Pads for Remote Center for Compliance (RCC). In: IEEE ICRA, pp. 291–298 (1996)
6. Havlik, S.: A New Elastic Structure for a Complaint Robot Wrist. Robotica. 1, 95–102 (1983)
7. Ciblak, N., Lipkin, H.: Design and Analysis of Remote Center of Compliant Structures. J. Robot. Syst. 20, 415–427 (2003)
8. Hricko, J., Havlík, S., Harťanský, R.: Optimization in Designing Compliant Robotic Micro-Devices. In: Int. Workshop on Robotics in Alpe-Adria-Danube Region – RAAD, Hungary, pp. 397–402 (2010)
9. Wang, M.Y.: A Kinetoelastic Formulation of Compliant Mechanism Optimization. ASME J. Mech. Robot. 1, 021011.1–021011.10 (2009)
10. Sigmund, O.: On the Design of Compliant Mechanisms using Topology Optimization. Mech. Struct. Mach. 25, 493–524 (1997)
11. Svanberg, K.: The Method of Moving Asymptotes: a New Method for Structural Optimization. Int. J. Numer. Meth. Eng. 24, 359–373 (1987)

Author Index

Printed in the United States
By Bookmasters